高等职业教育计算机专业系列教材

计算机专业英语

主　编　冯　于　蒋秀娟

副主编　郭汉英　李　敏

参　编　王瑞娇　吴婧聆

主　审　朱　丹　姚立佳

机 械 工 业 出 版 社

本书共12个单元，内容涵盖计算机硬件、操作系统、编程语言、网络等方面的专业英语知识。每个单元包括“听说（Listening & Speaking）”“阅读（Reading）”“计算机屏幕英语（Screen English）”3个部分。书后附录还提供了计算机专业英语词汇，便于学生使用和查询。另外，听力、对话练习以及课文均有配套录音，读者可以直接扫描对应二维码（建议在Wi-Fi环境下）进行播放，便于教学和自学。

本书一改传统计算机专业英语教材以阅读为主的模式，融听、说、读于一体，练习的设计也尽量做到“课证融通”，采用历年程序员、网络工程师考试中英语部分的真题，使学生在全面提高计算机专业英语应用能力的同时兼顾职业证书的获取。

本书不仅可以作为高职院校计算机相关专业的英语课程教材，也可作为广大计算机从业人员英语能力提升的自学用书。

为方便教学，本书配备电子课件等教学资源。凡选用本书作为教材的教师均可登录机械工业出版社教育服务网 www. cmpedu. com 免费下载。如有问题请致电 010－88379375 联系营销人员。

图书在版编目（CIP）数据

计算机专业英语／冯于，蒋秀娟主编. —北京：机械工业出版社，2016.5（2022.5 重印）

高等职业教育计算机专业系列教材

ISBN 978－7－111－53446－4

Ⅰ.①计… Ⅱ.①冯… ②蒋… Ⅲ.①电子计算机-英语-高等职业教育-教材 Ⅳ.①H31

中国版本图书馆 CIP 数据核字（2016）第067419号

机械工业出版社（北京市百万庄大街22号　邮政编码100037）
策划编辑：刘子峰　责任编辑：刘子峰
责任校对：刘子峰　封面设计：陈　沛
责任印制：常天培
河北鑫兆源印刷有限公司印刷
2022年5月第1版·第13次印刷
184mm×260mm·11.25印张·242千字
标准书号：ISBN 978－7－111－53446－4
定价：38.00元

电话服务	网络服务
客服电话：010-88361066	机　工　官　网：www. cmpbook. com
010-88379833	机　工　官　博：weibo. com/cmp1952
010-68326294	金　　书　　网：www. golden- book. com
封底无防伪标均为盗版	机工教育服务网：www. cmpedu. com

“计算机专业英语”是高职院校计算机专业的一门必修课程，目标是让学生掌握计算机专业英语词汇、科技应用英语的阅读及表达的方法和技巧，提高计算机方面的英文应用能力，从而在实际工作中受益。它要以公共英语课程的学习为基础，也是进一步学习计算机专业核心课程的基础。

本书是2010—2011年海南省高职院校英语类课程教学改革项目“计算机专业英语”（项目号：Hyjc2011－09）的研究成果，秉承“工学结合”的教育理念，采用“行业人员＋英语教师＋专业教师”的编写模式，取各方所长：行业人员提供计算机相关行业的工作过程、典型工作环节和场景；英语教师以所涉及场景为参照组织内容，根据主要工作任务所需的专业英语知识和技能设计计算机英语学习任务；专业教师对所有内容进行专业把关。本书不仅注重培养学生用英语处理与计算机工作相关的业务的能力，而且兼顾交际技能、职业技能和自主学习能力的培养。

本书在体例和内容上都有鲜明的特色，主要表现在以下几个方面：

1）听、说、读结合。本书遵循听、说、读结合的原则，以语言输出为目的，打破了传统计算机专业英语教材只重阅读、不重听说的编排格局。听力、对话练习以及课文均配有录音，读者可以直接扫描对应二维码（建议在Wi-Fi环境下）进行播放。

2）对话内容选材安排合理。本书选择了IT职场常见的话题，每个话题分别由英汉对照对话、英语词汇等组成，指导读者用英语处理实际工作，以此提高用英语解决实际问题的能力。另外，对话的编写简明、通俗、语言规范、表达流畅，读者可以有效地学到真正实用的英语表达方式。对话之后的任务设置为读者提供了进行交际活动所需的语言材料，旨在帮助读者学以致用。

3）体现“课证融通”。本书新增计算机相关考试英语部分全真试题与模拟题，为学生顺利通过专业考试提供帮助。

4）屏幕英语为学生实际操作排忧解难。屏幕英语的选取充分考虑到学生实际操作的需要，也有助于学生阅读计算机专业文档。

本书由冯于、蒋秀娟任主编，负责编写大纲和全书的统稿工作；郭汉英、李敏任副主编，负责协助主编的各项工作；参加编写的还有王瑞娇、吴婧聆。具体分工如下：冯于负责编写第3、11单元；蒋秀娟负责编写第1、10单元；郭汉英负责编写第6、8单元；李敏负责编写第 4、12 单元；王瑞娇负责编写第 2、7 单元；吴婧聆负责编写第 5、9 单元。朱丹、姚立佳对全部书稿进行了认真审阅。

本书在编写过程中，得到了海南软件职业技术学院教务处原处长陈鹤年教授以及海南软件职业技术学院软件工程系卢彧工程师的专业指导；教育部评估专家组成员、海南省大学外语研究会会长、海南师范大学外国语学院原院长陈宗华教授以及广东外语外贸大学英语语言文化学院教授、硕士生导师黄家修教授给本书的编写提出了宝贵意见；另外，海南软件职业技术学院的各级领导与同仁也给予了本书编者大力的支持与帮助，在此一并表示由衷的感谢！

由于编者水平有限，书中不足之处在所难免，恳请广大读者批评指正。

编　者

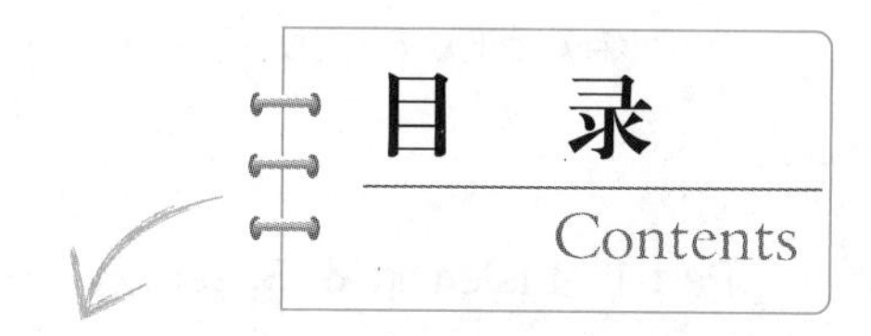

前言

Unit 1 Computer Hardware ······ 1
Part 1 Listening & Speaking ······ 2
Part 2 Reading ······ 5
Text Hardware ······ 5
Supplementary Reading Applying Technology ······ 11
Part 3 Screen English ······ 12
Key to Exercises ······ 13

Unit 2 Operating System ······ 15
Part 1 Listening & Speaking ······ 16
Part 2 Reading ······ 19
Text Operating System ······ 19
Supplementary Reading Operating System Types ······ 27
Part 3 Screen English ······ 28
Key to Exercises ······ 29

Unit 3 Programming Languages ······ 30
Part 1 Listening & Speaking ······ 31
Part 2 Reading ······ 33
Text Introduction to the History of Programming Languages ······ 33
Supplementary Reading Different Kinds of Programming Languages ······ 39
Part 3 Screen English ······ 42
Key to Exercises ······ 42

Unit 4 Office Automation ······ 44
Part 1 Listening & Speaking ······ 45
Part 2 Reading ······ 47
Text Office Automation ······ 47

Supplementary Reading　How to Write in Word? 52
Part 3　Screen English 53
Key to Exercises 55

Unit 5 **Browser** 56
Part 1　Listening & Speaking 57
Part 2　Reading 60
Text　Browser 60
Supplementary Reading　Some Popular Browsers 66
Part 3　Screen English 67
Key to Exercises 68

Unit 6 **Multimedia** 70
Part 1　Listening & Speaking 71
Part 2　Reading 73
Text　Multimedia 73
Supplementary Reading　Sound Card 79
Part 3　Screen English 81
Key to Exercises 81

Unit 7 **Search Engine** 83
Part 1　Listening & Speaking 84
Part 2　Reading 88
Text　Searching the Internet 88
Supplementary Reading　The Best Search Engines of 2011 95
Part 3　Screen English 97
Key to Exercises 98

Unit 8 **Internet** 100
Part 1　Listening & Speaking 101
Part 2　Reading 103
Text　The Internet Community Today 103
Supplementary Reading　Networks 108
Part 3　Screen English 110
Key to Exercises 111

Unit 9 **Tools for Online Communication** 112
Part 1　Listening & Speaking 113

Part 2 Reading ······ 116
Text Tools for Online Communication ······ 116
Supplementary Reading Comparing Wikis with Other Online Communication Tools ······ 123
Part 3 Screen English ······ 125
Key to Exercises ······ 126

Unit 10 Electronic Commerce ······ 128
Part 1 Listening & Speaking ······ 129
Part 2 Reading ······ 131
Text Electronic Commerce ······ 131
Supplementary Reading Taobao ······ 137
Part 3 Screen English ······ 139
Key to Exercises ······ 140

Unit 11 Install and Configure Software Programs ······ 141
Part 1 Listening & Speaking ······ 142
Part 2 Reading ······ 144
Text Instructions in Microsoft Word 2007: How to Install Microsoft Word 2007 ······ 144
Supplementary Reading How to Configure Virtual PC for USB Printing ······ 149
Part 3 Screen English ······ 150
Key to Exercises ······ 150

Unit 12 Computer Security ······ 152
Part 1 Listening & Speaking ······ 153
Part 2 Reading ······ 155
Text Computer Security ······ 155
Supplementary Reading What Is a Firewall? ······ 160
Part 3 Screen English ······ 162
Key to Exercises ······ 162

附 录 计算机专业英语词汇表 ······ 164

参考文献 ······ 172

Computer Hardware

Learning Objectives

After completing this unit, you will be able to:

1. Identify the four types of microcomputers;
2. Describe the different types of computer hardware, including input, output, storage and communication devices;
3. Explain the functions of the hardware components commonly found inside the system unit.

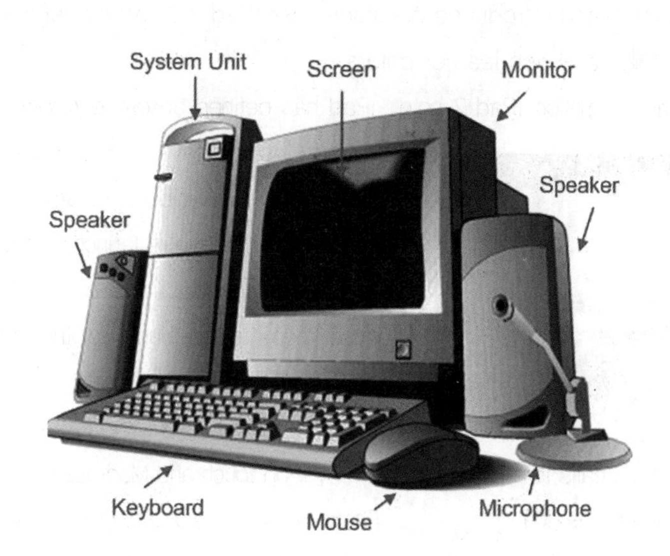

Part 1 Listening & Speaking

1. Listen to the following passage and fill in the blanks with the words in the box.

CEO	generation	introduced	email	visual	ebooks
screen	processor	battery	browsing	videos	dual-core
cameras	available	Wi-Fi	visual		

Apple today 1) ________ iPad 2, the next 2) ________ of its magical device for 3) ________ the Web, reading and sending 4) ________, enjoying photos, watching 5) ________, listening to music, playing games, reading 6) ________ and much more. IPad 2 features an entirely new design that is 33 percent thinner and up to 15 percent lighter than the original iPad, while maintaining the same stunning 9.7-inch LED-backlit LCD 7) ________. IPad 2 features Apple's new 8) ________ A5 9) ________ for blazing fast performance and stunning graphics and now includes two 10) ________, a front-facing VGA camera for FaceTime and Photo Booth, and a rear-facing camera that captures 720p HD video, bringing the innovative FaceTime feature to iPad users for the first time. Though it is thinner, lighter, faster and packed with new features, iPad 2 still delivers up to 10 hours of 11) ________ life that users have come to expect. iPad 2 is 12) ________ in black or white, features models that run on AT&T's and Verizon's 3G networks, and introduces the innovative IPad 2 Smart Cover in a range of vibrant polyurethane (聚氨酯) and rich leather colors.

"With more than 15 million IPad 2 sold, iPad has defined an entirely new category of mobile devices," said Steve Jobs, Apple's 13) ________. "While others have been scrambling to copy the first generation iPad, we're launching iPad 2, which moves the bar far ahead of the competition and will likely cause them to go back to the drawing boards yet again."

With the new front and rear cameras, IPad 2 users can now make FaceTime calls to millions of iPhone 4, iPod touch and Mac users so they can see family and friends anywhere there is 14) ________. Photo Booth lets you apply fun 15) ________ effects, including eight photo special effects like Squeeze, Twirl and Kaleidoscope, to photos captured by either camera.

2. Choose the proper words or expressions.

Have you ever wondered how information processed by the system unit is 1) ________ into a form that you can use? That is the role of output devices. While input devices convert what we understand into what the system unit can process, output devices convert what the system unit has processed into a form that we can understand. Output devices 2) ________ machine language into letters, numbers, sounds, and images that people can understand.

Competent end users need to know about the most commonly used input devices, including keyboards, mice, 3) ________, digital cameras, digitizing tablets, voice recognition, and MIDI devices. Additionally, they need to know about the most commonly used output devices. And end users need to be aware of 4) ________ input and output devices such as fax machines, multifunctional devices, 5) ________ telephones, and terminals.

1) A. converted B. convinced
2) A. transact B. translate
3) A. scanners B. scan
4) A. combination B. communication
5) A. Intranet B. Internet

3. Please read the conversation below and learn how to raise a question and how to solve each problem.

开机 start my computer	提示 prompt	丢失 miss
启动 boot	出错 go wrong	故障恢复控制台 the recovery console
修复 fix	应急修复 emergency repair	过程 process
启动盘 startup disk	按 press	安装程序 installation
命令提示框 command prompt	重启 restart	

A：马克，请帮帮我。我开机时，Windows 7 出故障了。系统提示"NTLDR 丢失"。

B：这说明系统启动所需的文件不存在或是出错了。

A：那我该怎么做呢？

B：你可以使用故障恢复控制台来修复这个问题。

A：我在哪里可以找到它呢？

B：你可以在应急修复的过程中运行它。使用 Windows 启动盘来启动计算机。当你看到"欢迎安装"信息时，按 R 键来"修复"。再按 C 键来运行故障恢复控制台。

A：之后我要做什么呢？

B：选择你想修复的安装程序并输入管理员密码。在故障恢复控制台的命令提示框中输入命令。然后重启你的计算机。

A：非常感谢。系统现在可以正常启动了。

B：不客气，有问题随时来找我。

A：Mark，I need your help. When I start my computer，Windows 7 won't work. And I was prompted "NTLDR is missing."

B：That means the boot files needed to start the system are missing or go wrong.

A：What should I do about this?

B：By using the Recovery Console，you can fix this problem.

A：Where can I find it?

B：You can run it from the Emergency Repair process. Use the Windows startup disk to start the computer. When you see the "Welcome to Setup" message，press R for "repair". Then press C to run the Recovery Console.

A：After that，what shall I do?

B：Select the installation that you want to repair and type in the administrator password. At the Recovery Console command prompt，type the commands. And then restart your computer.

A：Thank you very much. The system starts up properly now.

B：That's all right. If you have any question，don't hesitate to ask me.

4. Oral Practice

Mark 在自己的计算机上安装了不少软件，导致系统运行缓慢。请你告诉他应该如何把不必要的软件卸载掉。

Key Words

程序 program	控制面板 control panel	目录 directory
删除 delete	添加/删除 add/remove	文件夹 folder
卸载 uninstall		

Part 2 Reading

Hardware

Computers are electronic devices that can follow instructions to accept input, process that input, and produce information. This passage focuses principally on microcomputers. However, it is almost certain that you will come in contact, at least indirectly, with other types of computers.

Types of Computers

There are four types of computers: supercomputers[1], mainframe computers, minicomputers, and microcomputers.

Q1 How many types of computers are there? What are they?

Microcomputers are the least powerful, yet the most widely used and fastest-growing type of computer. There are four types of microcomputers: desktop, notebook, tablet PC, and handheld computers. Desktop computers are small enough to fit on top of or alongside a desk yet are too big to carry around. Notebook computers, also known as laptop computers, are portable, lightweight, and fit into most briefcases. A tablet PC is a type of notebook computer that accepts your handwriting. This input is digitized and converted to standard text that can be further processed by programs such as a word processor. Handheld computers are the smallest and are designed to fit into the palm of one's hand. Also known as palm computers, these systems typically combine pen input, writing recognition, personal organizational tools, and communications capabilities in a very small package. Personal digital assistants (PDAs) are the most widely used handheld computer.

Q2 How many types of microcomputers are there? What are they?

Microcomputer Hardware

Hardware for a microcomputer system consists of a variety of different devices. This physical equipment falls into four basic categories: system unit, input/output, secondary storage, and communication.

Q3 Can you describe the four basic categories of microcomputer hardware?

System unit: The system unit is a container that houses most of the electronic components that make up a computer system.[2] Two important components of the system unit are the microprocessor and memory. The microprocessor controls and manipulates data to produce information. Many times the microprocessor is contained within a protective cartridge. Memory, also known as primary storage or random access memory (RAM), holds data and program instructions for processing the data. It also holds the processed information before it is output. Memory is sometimes referred to as temporary storage because its contents will typically be lost if the electrical power to the computer is disrupted.

Q4 What are the two important components in the system unit?

Input/output: Input devices translate data and programs that humans can understand into a form that the computer can process. The most common input devices are the keyboard and the mouse. Output devices translate the processed information from the computer into a form that humans can understand. The most common output devices are monitors and printers.

Q5 What are the most common input and output devices?

Secondary storage: Unlike memory, secondary storage holds data and programs even after electrical power to the computer system has been turned off. The most important kinds of secondary media are hard and optical disks. Hard disks are typically used to store programs and very large data files. Using a rigid metallic platter, hard disks have a much greater capacity and are able to access information much faster than floppy disks. Optical discs use laser technology and have the greatest capacity. Three types of optical discs are compact discs (CDs), digital versatile (or video) discs (DVDs), and high-definition discs.

Communication: At one time, it was uncommon for a microcomputer system to communicate with other computer systems. Now, using communication devices, a microcomputer can communicate with other computer systems located as

Q6 What is the most widely used communication device? What is its function?

near as the next office or as far away as halfway around the world using the Internet.[3] The most widely used communication device is a modem, which modifies telephone communications into a form that can be processed by a computer.[4] Modems also modify computer output into a form that can be transmitted across standard telephone lines.

New Words

hardware ['hɑːdwɛə]	n. [计] 硬件 e. g. A computer system is composed of software and hardware in the light of its working mode. 从计算机的工作过程看，计算机系统由软件和硬件组成。
electronic [iˌlek'trɔnik]	a. 电子的，电子学的 e. g. The machine is operated by an electronic pulse. 这台机器由电子脉冲操控。
device [di'vais]	n. 设备，仪器，装置 e. g. The television receiver is an electronic device. 电视接收器是一种电子装置。
instruction [in'strʌkʃən]	n. (输入计算机的) 指令
briefcase ['briːfkeis]	n. 公文包
input ['input]	n. 投入，输入信息，输入数据，信息输入端 v. 输入 e. g. There are several errors in the input. 输入中有好几处错误。 I have inputted the data into a computer. 我已经把数据输入计算机了。
digitize ['didʒitaiz]	vt. [计] (将资料) 数字化 e. g. She bought device to digitize the data. 她买了能将数据数字化的装置。
recognition [ˌrekəg'niʃən]	n. 识别 e. g. His face was bruised and swollen almost beyond recognition. 他的脸又青又肿，几乎都认不出来了。
memory ['meməri]	n. 记忆力，回忆，记忆，[计] 存储 e. g. The computer has two million bytes of memory. 该计算机的存储器大小为两百万字节。

manipulate [mə'nipjuleit]	*vt*. 操纵，操作，控制，利用，处理，篡改 e. g. A clever politician know how to manipulate public opinion. 聪明的政客知道如何操纵公众舆论。
temporary ['tempəreri]	*a*. 临时的，暂时的，一时的 e. g. Ellen has got a temporary job. 艾伦找到一份临时工作。
data ['deitə]	*n*. 数据，资料 e. g. Have you recorded the data in disc? 你把那些数据存储到磁盘上了吗？
capacity [kə'pæsiti]	*n*. 容量
modify ['mɔdifai]	*vt*. 更改，修改 e. g. We have to modify our plan a little bit. 我们得对我们的计划稍加修改。
transmit [trænz'mit]	*vt*. 播送，发射，传送

Key Terms

microcomputer 微型计算机
supercomputer 巨型计算机
mainframe 主机，大型计算机
minicomputer 小型计算机
desktop 桌面，台式计算机
notebook 笔记本，便携式计算机
tablet PC 平板电脑
PDA (personal digital assistant) 个人数字助理
storage 贮藏，保管，存储
system unit [计] 系统处理单元
microprocessor 微处理器
keyboard 键盘
mouse 鼠标
monitor 显示器
printer 打印机
secondary storage 辅助存储器，二级存储器
modem 调制解调器

Useful Expressions

at least 至少

e. g. He smoked at least half a packet of cigarettes a day. 他每天至少抽半包烟。

consist of 由……组成

e. g. ... a computer-oriented language whose instructions consist only of computer instructions. ……一种面向计算机的语言，其指令只由计算机指令组成。

a variety of 种种，多种多样的

e. g. Computers come in a variety of sizes and shapes and with a variety of processing capabilities. 计算机的大小和形状不同，并且处理能力也各有差异。

refer to 涉及，指的是，提及，参考，适用于

e. g. Don't refer to this matter again, please. 请不要再提这件事了。

For further particulars, please refer to Chapter Ten. 详情请看第 10 章。

translate into 把……译成 把……转化成

e. g. ... a program to translate assembly language into computer language. ……把汇编语言翻译成计算机语言的程序。

communicate with 与……交流

e. g. How can a computer communicate with others? Channels and networks make it realizable. 计算机之间如何能够通信？这就要借助于通信信道和网络。

Notes

1. supercomputer

巨型计算机：一种超强数字计算机，通常指在指定时间应用的最快且高度准确的系统。巨型计算机通常有不止一个中央处理机，可同时进行工作。巨型计算机具有巨大的存储容量，非常快的输入/输出能力，能够同时处理大批数据的相应元素而不是每次处理一对元素。

2. The system unit is a container that houses most of the electronic components that make up a computer system.

该句子中有两个 that 引导的定语从句，第一个 that 引导的定语从句用来修饰 container，第二个 that 引导的定语从句用来修饰 electronic components。

3. Now, using communication devices, a microcomputer can communicate with other computer systems located as near as the next office or as far away as halfway around the world using the Internet.

在此句中 using communication devices 作方式状语，located as near as the next office or as far away as halfway around the world 过去分词作后置定语，using the Internet 作后置定语修饰 other computer systems.

4. The most widely used communication device is a modem, which modifies telephone communications into a form that can be processed by a computer.

此句中 which 引导非限制性定语从句，that 引导定语从句修饰 a form。

Exercises

1. Match each of the following terms with its Chinese equivalent.

1) microcomputer　　　　　　　　　　a. 微型计算机

2) tablet PC
3) desktop
4) PDA
5) microprocessor
6) keyboard
7) monitor
8) printer
9) modem
10) mouse

b. 调制解调器
c. 平板电脑
d. 打印机
e. 显示器
f. 台式计算机
g. 个人数字助理
h. 微处理器
i. 鼠标
j. 键盘

2. Recognize the following abbreviations by matching them with their corresponding full names and translate them into Chinese.

1) PDA ________________
2) PC ________________
3) CD ________________
4) DVD ________________
5) VCD ________________
6) Wi-Fi ________________
7) LED ________________
8) IT ________________
9) RAM ________________
10) CPU ________________

a. personal computer
b. compact disc
c. personal digital assistant
d. wireless-fidelity
e. information technology
f. light-emitting diode
g. random access memory
h. central processing unit
i. digital video disc
j. video compact disc

3. Complete the following sentences by translating the Chinese in the brackets.

1) Hardware for a microcomputer system ________________ (由各种不同的设备构成).

2) Microcomputers are the least powerful, yet the ________________ (使用最广且增长速度最快的) type of computer.

3) This input ________________ (被数字化并转换成标准文本) that can be further processed by programs such as a word processor.

4) The system unit is a container that ________________ (容纳大多数的电子元器件) that make up a computer system.

5) ________________ (最常见的输入设备) are the keyboard and the mouse.

4. Choose the best answer for each blank. (2008 年下半年程序员考试上午试题 (B))

1) As an operating system repeatedly allocates and frees storage space, many physically separated unused areas appear. This phenomenon is called ________ .

A. fragmentation B. compaction C. swapping D. paging

2) To document your code can increase program ________ and make program easier to ________.

A. reliability, execute B. security, interpret

C. readability, maintain D. usability, compile

3) We can use the word processor to ________ your documents.

A. edit B. compute C. translate D. unload

4) ________ infected computer may lose its data.

A. File B. Data base C. Virus D. Program

Supplementary Reading

Applying Technology

Webcam

Webcam is the input device commonly used to capture digital video to be displayed on the Web and to support video conferencing. This process typically involves four steps.

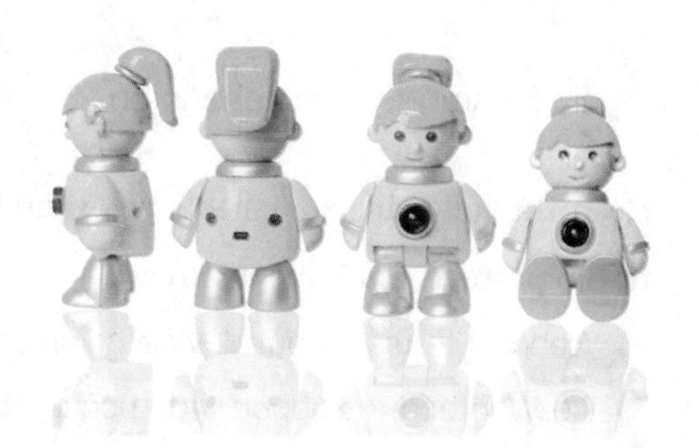

1. Input. A webcam continuously captures video and sends the images to your system unit.

2. Display. The video can be viewed directly on your monitor and/or over the Internet.

3. Web site. The video or select images can be posted to a Web site or Weblog for others to view.

4. Video conference. Webcam is often used to support video conferencing or the direct live face-to-face communications between individuals over the Internet.

Internet Telephony

Internet telephony is the transmission of telephone calls over computer networks. One of the most common applications involves computer-to-traditional telephone communications. This process uses a special Internet phone service provider and typically involves four steps.

1. Request. Using the software provided by the Internet phone service provider, the caller enters a telephone number and requests a connection.

2. Relay. The request is relayed to the provider's Internet server that is located closest to the requested number.

3. Connect. Using traditional local telephone communication lines, the server connects to the requested telephone.

4. Talk. The requested telephone rings, the party answers, and communication begins using the local telephone communication line and the Internet.

New Words and Expressions

demonstrate ['demənstreit] vt. 证明，演示，示范

webcam ['webˌkæm] n. 网络摄像头（= web camera）

capture ['kæptʃə] vt. 捕获，占领，夺取，吸引，(用照片等) 留存

Internet telephony 网络电话技术

request [ri'kwest] vt. & n. 请求，要求

relay [ri'lei] vt. 转播，转达，中继，接力，接替

server ['səːvə] n. 服务器

video conferencing 电视会议

weblog ['weblɔg] n. 网络日志，博客

Reading Comprehension

Read the following statements below and decide if they are true (T) or false (F) according to the passage you have just read.

1) Webcam is the output device commonly used to capture digital video to be displayed on the Web and to support video conferencing. (　　)

2) If you want to set up a video conferencing, the process involves four steps: input, display, Web site and video conference. (　　)

3) Internet telephony is the transmission of telephone calls over telephone line. (　　)

4) Internet telephony involves four steps: request, relay, connect and talk. (　　)

5) Request means using the software provided by the Internet phone service provider, the caller enters a telephone number and requests a connection. (　　)

Part 3 Screen English

开机屏幕提示

提示信息	含　义
Diskette Drive B: None	软盘驱动器 B：没有
Pri. Master Disk: 122880MB	第一主硬盘：122880MB

（续）

提示信息	含　义
Pri. Slave Disk：None	第一从硬盘：没有
Sec. Master Disk：None	第二主硬盘：没有
Sec. Slave Disk：CD ROM，Mode 2	第二从硬盘：CD ROM，模式 2
Display Type：EGA/VGA	显示类型：EGA/VGA
Serial Port（s）：3F8 2F8	串行端口：3F8 2F8
Parallel Port（s）：378	并行端口：378
EDO DRAM at Row：None	EDO 动态存储器位于行的位置：没有
SDRAM DRAM at Row：23	同步动态存储器位于行的位置：23

提示

在系统开机提示信息及 CMOS 参数设置中，大量出现名词短语及用缩略形式表示的专业术语，而非采用完整的英语句子的形式，这在计算机屏幕英语的阅读中，须重点加以留意。

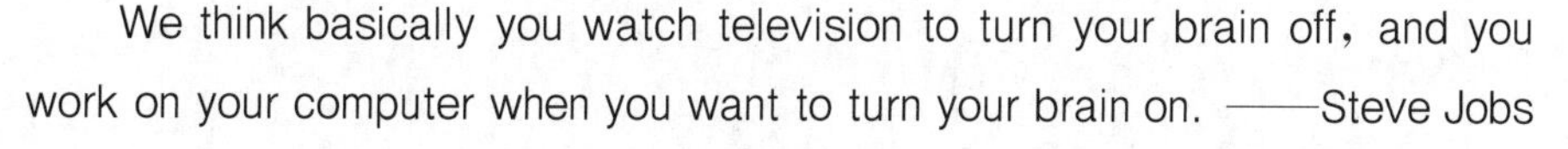

Quotation

We think basically you watch television to turn your brain off，and you work on your computer when you want to turn your brain on. ——Steve Jobs

我们认为看电视的时候，人的大脑基本停止工作，用计算机工作的时候，大脑才开始运转。

The only way to do great work is to love what you do. If you haven't found it yet，keep looking. Don't settle. As with all matters of the heart，you'll know when you find it.

—— Steve Jobs

成就一番伟业的唯一途径就是热爱自己的事业。如果你还没能找到让自己热爱的事业，继续寻找，不要放弃。跟随自己的心，总有一天你会找到的。

Key to Exercises

Listening & Speaking

1. 1）introduced　2）generation　3）browsing　4）email　5）videos
   6）ebooks　7）screen　8）dual-core　9）processor　10）cameras
   11）battery　12）available　13）CEO　14）Wi-Fi　15）visual
2. 1）A　2）B　3）A　4）A　5）B

Exercises

1. 1）a　2）c　3）f　4）g　5）h　6）j　7）e　8）d　9）b　10）i
2. 1）c 个人数字助理　2）a 个人计算机　3）b 光盘

4) i 数码光盘　5) j 视像光盘　6) d 无线保真（信号）
7) f 发光二极管　8) e 信息技术　9) g 随机存取存储器
10) h 中央处理器

3. 1) consists of a variety of different devices.
   2) most widely used and fastest-growing
   3) is digitized and converted to standard text
   4) houses most of the electronic components
   5) The most common input devices

4. 1) A　2) C　3) A　4) C

Comprehension

1) F　2) T　3) T　4) T　5) T

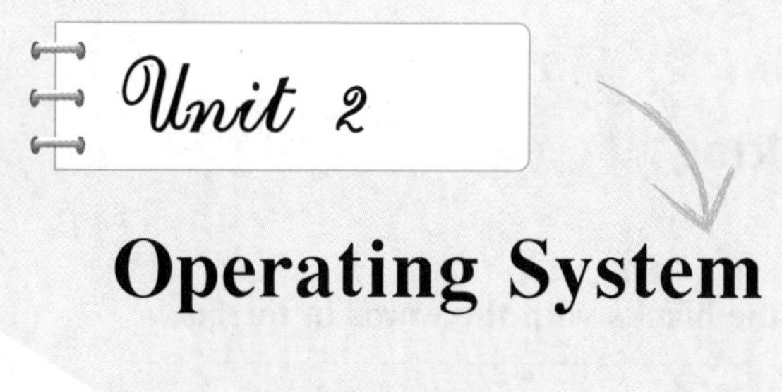

Operating System

Learning Objectives

After completing this unit, you will be able to:

1. Describe different types of operating system;
2. Explain the features of operating system;
3. Install an operating system;
4. Share photos online.

Windows

OS X

Linux

Part 1 Listening & Speaking

1. Listen to the following passage and fill in the blanks with the words in the box.

almost	platforms	share	debut	migration	same
alike	time	market	environment	four	dropped
individual					

Windows XP continues to shed market 1) ________ but still won't give up its crown as the most used operating system.

Eyeing the OS 2) ________ in October, Net Applications pegged the total share for XP at 48 percent, the first time the 10-year-old OS has dipped below 50 percent specifically among desktop 3) ________. That's down from 50. 5 percent in September and 52. 5 percent in August.

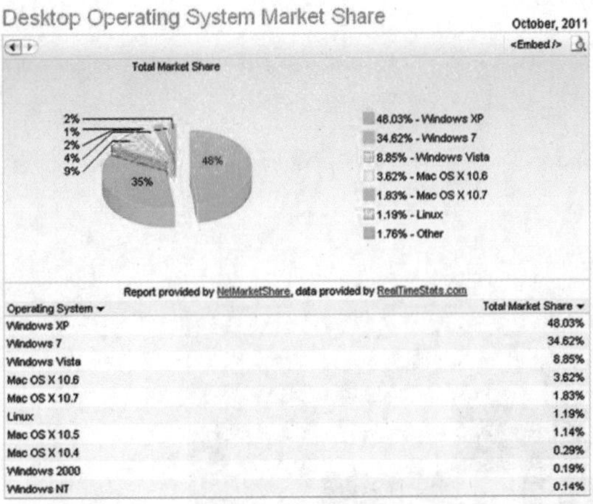

Desktop Operating System Market Share

October, 2011

Total Market Share

48.03% - Windows XP
34.62% - Windows 7
8.85% - Windows Vista
3.62% - Mac OS X 10.6
1.83% - Mac OS X 10.7
1.19% - Linux
1.76% - Other

Report provided by NetMarketShare, data provided by RealTimeStats.com

Operating System	Total Market Share
Windows XP	48.03%
Windows 7	34.62%
Windows Vista	8.85%
Mac OS X 10.6	3.62%
Mac OS X 10.7	1.83%
Linux	1.19%
Mac OS X 10.5	1.14%
Mac OS X 10.4	0.29%
Windows 2000	0.19%
Windows NT	0.14%

Since Windows 7 made its official 4) ________ more than two years ago, XP has gradually 5) ________ in usage from 72 percent in August 2009 to its current share. On a natural upswing, Windows 7 has grown over the 6) ________ stretch to capture almost 35 percent of the market as of last month. At the same time, Windows Vista's popularity has dropped to just under 9 percent.

Microsoft has aggressively been trying to convince consumers and corporations 7) ________ to upgrade from XP to Windows 7. The company keeps pushing the message that support for XP will end in April 2014. And though that sounds like a distant date for the 8) ________ user, enterprises typically need a healthy amount of 9) ________, money, and resources to plan and implement a major OS 10) ________.

Microsoft even told businesses still on XP not to wait for Windows 8 and forge ahead with a migration to Windows 7. The company has also been touting Windows 7 as a more secure 11) ________ than Windows XP. Its May Security Intelligence Report found that Windows 7 is 12) ________ to five times less vulnerable to malware infections than XP is.

Looking at the OS arena in general, Net Applications found that Windows accounted for 13) ________ 92 percent of the market in October, while the Mac OS took home almost 7 percent.

2. Choose the proper words or expressions.

San Francisco, May 24. Microsoft Corp. on Tuesday officially unveiled the latest version of its mobile operating system, touting new features that differentiate itself from competitors.

The major new update of Microsoft's Windows phone software, code named Mango, includes hundreds of new features that will 1) ________ smarter and easier communications, applications and Internet experiences, the company said.

For example, the new software can 2) ________ a user's text message, email, Facebook and Twitter chats, Windows Live Messenger in one, easy-to-access location, allowing the user to switch between them within the same conversation.

There is also a 3) ________ that can connect applications already on a Windows phone, or new applications available to download, with search results and users' other activities in a way that Microsoft said is 4) ________ than any other platform.

Mango will use a mobile version of Microsoft's Internet Explorer 9 browser, and will leverage built-in phone capabilities like location awareness, camera and microphone to offer more relevant search results and local information and suggestions.

"Mango builds on the work that we did in Windows Phone 7 and extends a lot of key scenarios around communications, apps, and Internet experiences with even more capability and a deeper level of integration," Greg Sullivan, senior product manager of mobile communications at Microsoft, said in a statement.

Microsoft said Mango will be offered for free to all eligible Windows phone customers when it's available in the fall.

The company on Tuesday also announced new partnerships with Acer, Fujitsu and ZTE in its latest efforts to bring more Windows phone 5) ________ to the market.

1) A. deliver　　B. delay

2) A. integrant　　B. integrate

3) A. feature　　B. future

4) A. deepen　　B. deeper

5) A. handsets　　B. handbook

3. Please read the conversation below and learn how to raise a question and how to solve each problem.

安装 install	用途 usage
人性化 humane	容易做某事 be easy to do something
特性 feature	硬件配置 hardware configuration
操作系统 operating system	修改 revise
文件 file	

A：汤姆，我刚买了台计算机，想安装个操作系统，你认为哪个比较好呢？或者你认为哪个版本的好些呢？

B：那要看你的计算机的用途。

A：系统有很多种吗？

B：是的，主要有两种，一种是 Windows 系统的，另一种是 Linux 系统的。每一种都有个人版和专业版等版本。

A：个人版应该比较适合我吧。

B：如果就你自己使用，是的。我认为 Windows 的适合你。

A：为什么呢？

B：与 Linux 的相比，Windows 的操作系统界面比较人性化，对计算机初学者来说比较容易操作使用，而且网上有许多与之相应的资源，比如 Microsoft Office、MSN，IE 等。

A：哦，我明白了。那 Windows 的最新版是什么？

B：Windows 7，它有许多特性，但是对硬件要求比较高。可以让我看看你的硬件配置吗？

A：当然！

B：CPU……没问题。

B：那赶紧帮我装上吧。记得要帮我装上 Microsoft Office，我还要修改一份文件呢。

A：好的。

A：Tom，I've just bought a computer，and I want to install an operating system. Can you give me some advice? Or which version do you think is better?

B：That depends on the usage of your computer.

A：Are there many kinds of operating systems?

B：Yes，generally speaking，there are two. One is Windows'，and the other is Linux's. Each of them has Home Basic，Professional，etc.

A：Maybe Home Basic is good for me.

B：If you use the computer only by yourself，the answer is "Yes". And Windows' is a good choice.

A：Why?

B：Compared with Linux's，Windows' Operating System Interface is humane. For a green hand，it's easy to operate. In addition，there are many corresponding online resources，such as Microsoft Office，MSN，Internet Explorer，etc.

A：Oh，I see. What's the latest version of Windows?

B：Windows 7. It has a lot of features，but relatively requires better hardware. Can I have a look at your hardware configuration?

A：Of course.

B: CPU... No problem.

A: Please help me install it. Don't forget to install Microsoft Office; I need to revise a file.

B: It's my pleasure.

4. Oral Practice

Mark 想在网上和家人共享照片。请你告诉他应该怎么做。

Key Words

共享 share	家庭组 homegroup	控制面板 control panel
点击 click	选择 select	密码 password
接入 access	"开始" 按钮 Start button	

Part 2 Reading

Text

Operating System

The most essential program on any computer is the operating system or OS. It is a composition of many smaller programs to control the CPU. The operating system performs basic tasks which make hardware devices work together, such as identifying keyboard input, sending output to the screen, tracking files stored on disk and directory, and controlling peripheral devices such as disk drives and printers. In other words, the operating system makes computers easier to use.

Q1 What is an operating system?

Types of Operating System

Operating systems are mainly divided into three categories: embedded, network, and stand-alone operating systems.

Q2 How many types are operating systems categorized into? What are they?

The embedded operating system is applied to handheld computers, like PDAs. This operating system is stored within the ROM memory. Some popular embedded operating systems are Windows CE[1] and Palm OS.

Q3 Where is the operating system used?

The network operating system is aimed to run the computers which are associated together via a network. This operating system is typically placed on the hard disk of one of the linked computers. This computer is known as a network server, which has to adjust the communication between networked computers. Some popular network operating systems are Netware, Windows NT Server, Windows XP Server, and UNIX.

Q4 Can you list some popular network operating systems?

The stand-alone operating system, also named the desktop operating system, is situated on the desktop and notebook computer's hard disks to control the computer operations. Popular desktop operating systems include Windows, Mac OS, and UNIX.

Q5 Where are stand-alone operating systems located?

To the operating system, ROM and RAM are also very significant. Part of a computer's operating system is built into ROM. The ROM operating system is also identified as the BIOS (basic input output system). The BIOS is responsible for waking up the computer when you boot it. It also reminds the computer of all the parts it has and what they do.[2]

The operating system also includes other important programs. This part of the operating system which is stored on a computer's hard drive is known as DOS. When the computer is booted, it is read to RAM.

Functions of Operating System

Although the manufacturers of operational software vary, they all have similar characteristics. Modern hardware, because of its sophistication, requires that operating system meets some certain specific standards.[3] For example, taking into account the current state of the field, an operating system must support some form of online processing. Functions usually related with operational software are:

Q6 What are the functions of operating system?

1. Resource Management

The resource management in a computer system is another key aspect of the operating system. Obviously, a program cannot use a device if that hardware is unavailable. As we have seen, the operational software oversees the execution of all programs. It also monitors the running devices. To achieve this, the program is being used or its equipment will be used to match the non-operating system by checking the table to allow or deny use of certain equipment.

2. Control of I/O Operations

Allocation of a system's resources is closely related to the operational software's control of I/O operations. Since it is very necessary to access a particular device before I/O operations may start, the operating system must organize I/O operation and devices on which they are performed. In fact, it sets up a directory of programs undergoing execution and the devices they must use in achieving I/O operations. Using control statements, jobs may require specific equipments which let users read data from exact sites or print information at selected offices. Taking advantage of this facility, data read from one site may be distributed throughout computerized.

Q7 Is the allocation of a system's resources closely connected to the operational software's control of I/O operations?

In order to facilitate the performance of I/O operations, most operating systems have a standard set of control commands to handle the processing of all input and output commands. In effect, in using a specific I/O device, the program in the implementation process of the system sends the command that an I/O operation is desired. The controlling software calls on the input/output control system software to actually perform the I/O operation. Considering the level of I/O activity in most programs, the IOCS (input/output control system) instructions are really vital.

3. Job Management

A very important responsibility of any operational software is the scheduling of jobs to be handled by a computer system which is one of the main tasks of the job management function. The operating system establishes the order in which programs are processed, and defines the sequence in which particular jobs are executed. The operating system weighs many factors in creating the job queue. Operating system establishes the procedure order, and defines a specific order of execution of the job. Operating software should be able to assess these factors and control the processing order.

New Words

perform [pə'fɔːm]	*vi*. 执行，完成 e. g. The system is inadequate for the tasks it has to perform. 这个系统不足以完成它的任务。
display [di'splei]	*n*. 显示，显示器 e. g. The addition of a display on your remote may also tempt you to allow navigation to multiple "pages" of content or controls on the display. 遥控器加上了这个小显示屏，让你可以通过它导航至多个页面或者控制功能。
track [træk]	*n*. 行踪，轨道，足迹 e. g. For most applications, this is quite feasible. Another approach is to keep track of small changes in memory and write them to the disk at reasonable intervals. 对于大多数程序来说，这是可行的。另一种尝试是记录内存中的细小变化，并在合理的时间间隔内将它们写入硬盘中。
via ['vaiə]	*prep*. 经过，通过，凭借，取道 e. g. If a clerk enters something erroneous, he needs to be informed of it via both auditory and visual feedback. 如果职员输入了一些错误数据，应该给他听觉和视觉两种反馈。
computerize [kəm'pjuːtəraiz]	*vt*. 使计算机化，用计算机处理 e. g. The firm decided to computerize its wages department. 公司决定用计算机管理发薪部门的工作。
peripheral [pə'rifərəl]	*a*. 周围的，外面的 e. g. Programmers like wizards because they get to treat users like peripheral devices. 程序员喜欢向导，因为它们像对待外围设备一样对待用户。
stand-alone	*a*. 单机的，独立的 e. g. Some are stand-alone clients, but many use a browser for display. 有些是单机的客户端，但很多用浏览器来演示。
embedded [im'bedid]	*a*. 嵌入的 e. g. Another distinct difference between embedded systems and desktop applications is the importance of environmental context. 嵌入式系统和桌面应用的另一个主要区别是使用情境的不同。

typically ['tipikli]	*ad*. 代表性地，作为特色地，典型地 e. g. Typically, a user clicks on an object and it becomes selected. 比如，用户单击一个对象，对象即被选中。 Typically, when a selection is made, any previous selection is unmade. 通常，选中一个对象后，之前选中的对象都不再被选中了。
located [ləu'keitid]	*a*. 坐落的，位于的 e. g. Are items located in the places where customers might find them? 物品是否都在顾客可能找到的地方？
coordinate [kəu'ɔːdneit]	*vt*. 协调，调节，使调和 e. g. Later in the chapter we will discuss how to coordinate interaction design with industrial design. 在本章后面，我们将讨论如何协调交互设计和工业设计。
manufacturer [ˌmænju'fæktʃərə]	*n*. 制造商
desktop ['desktɔp]	*n*. 台式计算机 e. g. Desktop applications fit into three categories of posture: sovereign, transient, and daemonic. 桌面应用有3种状态：独占、暂时和后台。
operational [ˌɔpə'reiʃənəl]	*a*. 操作的，运作的 e. g. Is the equipment operational yet? 这套设备投入运转了吗？
boot [buːt]	*vt*. 启动 e. g. Boot from CD: disk boot failure, insert system disk and press Enter. 从光驱启动：系统启动失败，请插入系统盘后，按"回车"键。
characteristics [ˌkæriktə'ristik]	*n*. 特性，特征，特色 e. g. Brands indicate the positive characteristics of the product and suggest discrimination and taste of the user. 品牌显示出产品的正面特征，体现着与其他产品的区别，也体现了用户的品位。
sophistication [səˌfisti'keiʃən]	*n*. 复杂，精密 e. g. He carried out this prerequisite with skill and sophistication. 他用自己的技巧和细心创造了这一前提条件。
associate [ə'səuʃieit]	*vt*. 使联合，使结合 e. g. We have never given our users a chance to experience high-quality, positive audible feedback in our programs, so it's no wonder that people associate sound with bad interfaces. 在程序中我们从来没有为用户提供过高质量的积极的听觉反馈，难怪人们会把声音与不好的界面联系在一起。

directory [di'rektəri]	n. 目录 e. g. Alternatively, he could look the address up in a directory. 还有一种办法，他可能会到目录中查找地址。
facility [fə'siliti]	n. 设备，设施 e. g. If the undo facility is sufficiently easy to use and understand, users won't be bothered by it. 如果撤销工具足够易用易懂，用户就不会被它所困扰了。
execution [ˌeksi'kjuːʃən]	n. 执行 e. g. ... a parameter that is bound during the execution of a computer program. ……一种在计算机程序执行期间被赋值的参数。
allocation [ˌælə'keiʃən]	n. 分配，定位 e. g. ... a system task that handles system initialization, allocation of time-shared regions, swapping, and general control of the time-sharing operation. ……一种能处理系统初始化、分时区域分配、交换和全面控制分时操作的系统任务。
unavailable [ˌʌnə'veiləbl]	a. 没有效果的，无用的 e. g. The repositioning function demands the click-and-drag action, making it unavailable for other purposes. 调整位置这一功能需要使用单击拖动操作，结果是这个操作不能再用于其他目的。
facilitate [fə'siliteit]	vt. 促进，帮助 e. g. Our trade and development program will facilitate our progress. 我们的贸易和发展计划将有利于我们的进步。
vital ['vaitl]	a. 极其重要的，必不可少的 e. g. When an object can be either selected or dragged, it is vital that the mouse be biased towards the selection operation. 当一个对象既可能被选择，也可能被拖动时，鼠标区偏向于选择操作，这非常关键。
schedule ['skedʒul]	vt. 将……列表（或清单等） e. g. George always falls behind his schedule. 乔治老是不能按时完成计划。

Key Terms

operating system 操作系统
hardware device 硬件设备
ROM (read-only memory) 只读存储器
ROM memory 只读存储器的内存
RAM (random access memory) 随机存储器
resource management 资源管理

Useful Expressions

be responsible for 对……负责

e. g. They all desire you to be responsible for the project. 他们都想让你负责这个项目。

vary with 随……而变化

e. g. Prices usually vary with the quality. 价格常随品质而变化。

be associated with 与……有关，与……有关系

e. g. Name of document class is to be associated with this document template. 文档类的名称要与此文档模板关联。

take advantage of 利用

e. g. I should like to take advantage of this opportunity to express my thanks for your help. 我想借这个机会，对你们的帮助表示感谢。

Notes

1. Windows CE

Windows CE是微软公司嵌入式、移动计算平台的基础，它是一个开放的、可升级的32位嵌入式操作系统，是基于掌上型计算机的电子设备操作系统。Windows CE的图形用户界面相当出色。CE中的C代表袖珍（compact）、消费（consumer）、通信能力（connectivity）和伴侣（companion）；E代表电子产品（electronics）。

2. It also reminds the computer of all the parts it has and what they do.

在此句中it has是修饰parts的定语从句，省略了关系代词that。

3. Modern hardware, because of its sophistication, requires that operating system meets some certain specific standards.

该句子中because of its sophistication做状语，表原因；that引导的是宾语从句，在此无意义，可省略。

Exercises

1. Match each of the following terms with its Chinese equivalent.

1) display	a. 资源管理
2) manufacturer	b. 独立操作系统
3) allocation	c. 嵌入式操作系统
4) embedded operating system	d. 作业管理
5) handheld computer	e. 生产商
6) stand-alone operating system	f. 显示器
7) network operating system	g. 掌上型计算机
8) job management	h. 输入输出操作控制

9) resource management　　i. 分配，定位
10) control of I/O operations　　j. 执行
11) execution　　k. 复杂，精密
12) sophistication　　l. 网络操作系统

2. Recognize the following abbreviations by matching them with their corresponding full names and translate them into Chinese.

1) OS ________　　a. input/output control system
2) BIOS ________　　b. read only memory
3) RAM ________　　c. random access memory
4) ROM ________　　d. basic input output system
5) DOS ________　　e. disk operating system
6) IOCS ________　　f. operating system
7) MPEG ________　　g. joint photographic experts group
8) JPEG ________　　h. graphical user interface
9) GUI ________　　i. motion picture experts group

3. Complete the following sentences by translating the Chinese in the brackets.

1) The network operating system ________________ (被设计用于控制计算机) that are linked together via a network.

2) The BIOS ________________ (负责唤醒计算机) when you turn it on.

3) ________________ (正如我们所看到的), the operational software oversees the execution of all programs.

4) The operating system sets up ________________ (程序处理的顺序), and defines the sequence in which particular jobs are executed.

5) ________________ (为便于 I/O 操作的执行), most operating systems have a standard set of control instructions to handle the processing of all input and output instructions.

4. Choose the best answer for each blank. (2008 年下半年程序员考试上午试题 (B))

1) One use of network is to let several computers share ________ such as file system, printers, and tape drives.

A. CPU　　B. memory　　C. resources　　D. data

2) A firewall is a ________ system designed to ________ an organization's network against threats.

A. operating, prevent　　B. programming, develop
C. security, protect　　D. service, exploit

3) The ________ has several major components, including the system kernel, a memory management system, the file system manager, device drivers, and the system libraries.

A. application B. information system

C. network D. operating system

4) ________ is the address of a variable or a variable in which the address of another variable is stored.

A. Director B. Pointer C. Array D. Record

5) C++ is used with proper ________ design techniques.

A. object-oriented B. object-based C. face to object D. face to target

Supplementary Reading

Operating System Types

An operating system, or OS, is a software program that enables the computer hardware to communicate and operate with the computer software. Without a computer operating system, a computer would be useless.

Computers have progressed and developed, so have the operating systems. Below is a basic list of the different operating systems and a few examples of operating systems that fall into each of the categories. Many computer operating systems will fall into more than one of the below categories.

GUI: Short for graphical user interface. A GUI operating system contains graphics and icons and is commonly navigated by using a computer mouse.

Multi-user: A multi-user operating system allows for multiple users to use the same computer at the same time and different times. See the multi-user definition for a complete definition for a complete definition.

Multi-processing: An operating system that is capable of supporting and utilizing more than one computer processor.

Multi-tasking: An operating system that is capable of allowing multiple software processes to run at the same time.

Multi-threading: An operating system that allows different parts of a software program to run concurrently.

New Words and Expressions

category [ˈkætigəri] *n*. 种类

graphical [ˈgræfikəl] *a*. 图像的

navigate [ˈnæviˌgeit] *v*. 定位

multiple [ˈmʌltipl] *a*. 多人享有（或参加）的

utilize [ˈjuːtilaiz] *vt*. 利用

concurrently [kənˈkʌrəntli] *ad*. 同时发生地，并存地

Reading Comprehension

Read the following statements below and decide if they are true (T) or false (F) according to the passage you have just read.

1) There are six types of operating systems mentioned here. (　　)

2) A GUI operating system contains graphics and icons and is commonly navigated by using a computer mouse. (　　)

3) Linux is one of the examples of multi-user operating systems. (　　)

4) Multi-tasking operating system is capable of supporting and utilizing more than one computer processor. (　　)

5) The operating system of multi-threading allows different parts of a software program to run at the same time. (　　)

Part 3 Screen English

BIOS 特性设置

项　目	参　数
Virus Warning　病毒报警	Enabled/Disabled① 打开/关闭
CPU Internal Cache　CPU 内部缓	Enabled/Disabled　启用/禁止
External Cache　CPU 外部缓	Enabled/Disabled　启用/禁止
Quick Power On Self Test　快速通电自检	Enabled/Disabled　允许/禁止
Boot Sequence　启动顺序	A，C，SCSI　按 A，C，SCSI 顺序启动系统
Swap Floppy Drive　交换软驱	Enabled/Disabled　交换/正常
Boot Up Floppy Seek　开机启动时是否自动检测软驱	Enabled/Disabled　测试/跳过测试
Boot Up Numlock Status　开机后数字小键盘状态	On/Off　数字输入状态/箭头输入状态
Typematic Rate (Chars/Sec)　击键速率设置（字符/秒）	6～30
Typematic Delay (Msec)　击键延迟时间（毫秒）	250～1000/250～1000
Security Option　安全选项	System/Setup　系统开机口令/设置口令
Video BIOS Shadow　视频 BIOS 映射	Enabled/Disabled　允许/禁止
Delay For HDD (Secs)　硬盘延迟（秒）	0～15/0～15

注①："Enabled/Disabled"参数在不同的设置选项中，相应的含义有所区别。可以表

示“允许/禁止”“打开/关闭”“启用/不启用”“设置/不设置”等含义，须重点加以留意。

Quotation

Be nice to nerds. Chances are you'll end up working for one.

—— Bill Gates

善待书呆子。有可能到头来你会为一个书呆子工作。

Television is NOT real life. In real life, people actually have to leave the coffee shop and go to jobs.

—— Bill Gates

电视并不是真实的生活。在现实生活中，人们实际上得离开咖啡屋去干自己的工作。

Key to Exercises

Listening & Speaking

1. 1) share　2) market　3) platforms　4) debut　5) dropped
   6) same　7) alike　8) individual　9) time　10) migration
   11) environment　12) four　13) almost
2. 1) A　2) B　3) A　4) B　5) A

Exercises

1. 1) f　2) e　3) i　4) c　5) g　6) b
   7) l　8) d　9) a　10) h　11) j　12) k
2. 1) f 操作系统　2) d 基本输入输出系统　3) c 随机存储器
   4) b 只读存储器　5) e 磁盘操作系统　6) a 输入输出控制系统
   7) i 活动图像专家组　8) g 图像专家联合小组　9) h 图形用户界面
3. 1) is designed to control the computers
   2) is responsible for waking up the computer
   3) As we have seen
   4) the order in which programs are processed
   5) To facilitate execution of I/O operations
4. 1) C　2) C　3) B　4) D　5) A

Comprehension

1) F　2) T　3) T　4) F　5) T

Programming Languages

Learning Objectives

After completing this unit, you will be able to:

1. Improve your Listening comprehension and oral English;
2. Understand the development of programming languages;
3. Make clear the differences between machine language and assembly language.

Part 1 Listening & Speaking

1. Listen to the following passage and fill in the blanks with the words in the box.

logic programming	imperative	functional programming	
instructions	invention of the computer	machine	programs

A programming language is an artificial language designed to communicate 1) ________ to a 2) ________, particularly a computer. Programming languages can be used to create 3) ________ that control the behavior of a machine and/or to express algorithms precisely.

The earliest programming languages predate the 4) ________, and were used to direct the behavior of machines such as Jacquard looms and player pianos. Thousands of different programming languages have been created, mainly in the computer field, with many more being created every year. Most programming languages describe computation in an 5) ________ style, i.e., as a sequence of commands, although some languages, such as those that support 6) ________ or 7) ________, use alternative forms of description.

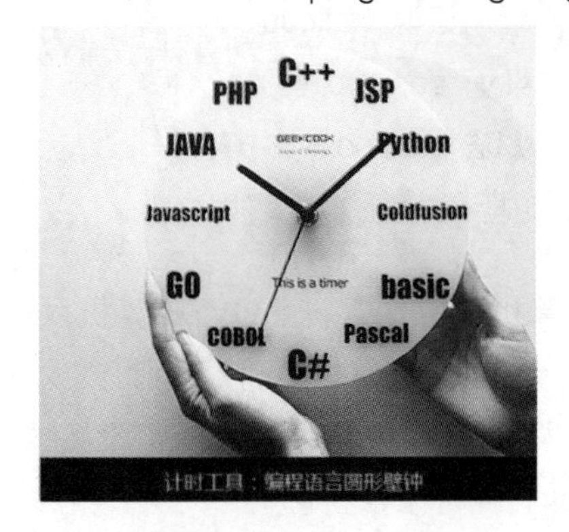

计时工具：编程语言圆形壁钟

2. Choose the proper words or expressions.

ABC is an interactive programming language and environment for personal computing, originally 1) ________ as a good replacement for BASIC. It was designed by 2) ________ a task analysis of the programming task.

ABC is easy to learn (an hour or so for someone who has already programmed), and yet easy to use. Originally intended as a language for 3) ________, it has evolved into a powerful tool for beginners and experts alike.

Some features of the language:

- a powerful collection of only 5 data types that can 4) ________ ;
- strong typing, yet without declarations;
- no limitations (such as max int), apart from sheer exhaustion of memory;
- refinements to support top-down programming;
- nesting by indentation;
- programs typically one fourth or one fifth the size of the equivalent Pascal or C.

Some features of the environment:

- no need for files: procedures and functions and global variables remain after logging out;
- one consistent face 5) ________ the user at all times, whether executing commands, editing, or entering input to a program;
- generalized undo mechanism.

1) A. intending　　B. intended

2) A. first does　　B. first doing

3) A. beginners　　B. beginning

4) A. easily be combined　　B. easy be combined

5) A. is shown to　　B. is showing to

3. Please read the conversation below and learn how to raise a question and how to solve each problem.

替代 replace	重新配置 reconfigure	单调乏味 tedious
转移 transfer	向导 wizard	桌面 desktop
显示 display	拨号 dial-up	方法 method
软盘 floppy disk	可移动媒介 removable media	项目 item

A: 托尼，我刚买了一台新的便携式计算机替代我的旧计算机。但是重新配置新计算机太单调乏味了。我怎么才能将旧计算机中的文件和设置转移到新计算机中呢?

B: 你可以使用 Windows XP 中的“文件和设置转移向导”来转移桌面设置、显示设置、拨号连接和其他类型的设置。单击“附件”“系统工具”就可以找到它。

A: 具体该怎么做呢?

B: 首先，你要从旧计算机中获得文件和设置信息。在旧计算机的向导里，第一步，选择“这是旧计算机”；第二步，选择一种转移方法，如软盘或其他可移动媒介；第三步，选择你需要转移的项目的相关选项。你可以从“仅设置”“仅文件”和“文件和设置”中选择一个。最后等待向导处理相关项目。

A: 在新计算机中我该如何操作呢?

B: 将文件和设置转移到新计算机中，在向导中选择“这是新计算机”。然后单击“我不需要向导磁盘。我已经从旧计算机中收集了我的文件和设置。”然后选择转移方法。

A: 非常谢谢，托尼。

B: 不客气。

A: Tony, I just bought a new laptop to replace my old one, but it is tedious to reconfigure the computer. How can I transfer files and settings from the old computer to the new one?

B: You can do that by using the "Files and Settings Transfer Wizard" in Windows XP. You can use this wizard to transfer desktop settings, display settings, dial-up connections, and other types of settings. You can find it by clicking Accessories, System Tools.

A: Can you tell me more about it?

B: First, you should get the files and the settings from the old computer. On the old computer, select Old Computer in the first step. The second step, select a transfer method, such as floppy drive or other removable media. The third step, click the option that corresponds to the items that you want to transfer. You can choose Settings Only, Files Only, or Both Files And Settings. At last, wait as the Wizard processes the items.

A: Well, what should I do on the new computer?

B: To transfer the files and the settings to the new computer, select New Computer in the Wizard. Then click "I don't need the Wizard Disk. I have already collected my files and settings from my old computer." And then select the transfer method.

A: Thank you very much, Tony.

B: You are welcome.

4. Oral Practice

John 不知道怎么给自己的文档上端加上页码，于是他虚心向 Alice 寻求帮助。

Key Words

菜单栏 menu bar　　视图 view　　页眉和页脚 header and footer

插入 insert　　光标 cursor

Part 2 Reading

Text

Introduction to the History of Programming Languages

The development of programming languages, unsurprisingly, follows closely the development of the physical and electronic processes used in today's computers.

In the 1940s the first recognizably modern, electrically powered computers were created, requiring programmers to operate machines by hand. Some military calculation needs were a driving force in early computer development, such as encryption, decryption, trajectory calculation and massive number crunching needed in the development of atomic bombs. At that time, computers were extremely large, slow and expensive. The advances in electronic technology in the post-war years led to the construction of more practical electronic computers. At that time only Konrad Zuse[1] imagined the use of a programming language (developed eventually as Plankalkul[2]) like those of today for solving problems.

Q1 What led to the construction of more practical electronic computers?

Q2 Who had the idea of the use of a programming language?

Subsequent breakthroughs in electronic technology (transistors, integrated circuits, and chips) dove the development of a variety of standardized computer languages to run on them. The improved availability and ease of use of computers led to a much wider circle of people who can deal with computers. The subsequent explosive development has resulted in the Internet, the ubiquity of personal computers, and increased use of computer programming, through more accessible languages such as Python[3], Visual Basic[4].

Q3 What made more and more people deal with computers?

A primary purpose of programming languages is to enable programmers to express their intent for a computation more easily than they could with a lower-level language or machine code. For this reason, programming languages are generally designed to use a higher-level syntax, which can be easily communicated and understood by human programmers. Programming languages are important tools for helping software engineers write better programs faster.

Q4 What mainly causes programming languages to use higher-level syntax?

Understanding programming languages is crucial for those engaged in computer science because today all types of computation are done with computer languages.

During the last few decades, a large number of computer languages have been introduced, have replaced each other, and have been modified/combined. Although there have been several attempts to make a universal computer language that serves all purposes, all of them have failed. The need for a significant range of computer languages is caused by the fact that the purpose of programming languages varies from commercial software development to hobby use; the gap in skill between novices and experts is huge and some languages are too difficult for beginners to come to grip with; computer programmers have different preferences; and finally, acceptable runtime cost may be very different for programs running on a microcontroller and programs running on a supercomputer.

Q5 Is there a programming language that can serve all purposes? Why or why not?

There are many special purpose languages, for use in special situations: PHP[5] is a scripting language that is especially suited for Web development; Perl[6] is suitable for text manipulation; the C language that has been widely used for development of operating systems and compilers (so-called system programming).

Q6 What special purpose languages have been listed in this text?

Programming languages make computer programs less dependent on particular machines or environments. This is because programming languages are converted into specific machine code for a particular machine rather than being expected directly by the machine. One ambitious goal of FORTRAN[7], one of the first programming languages, was this machine-independence.

New Words

military [ˈmilitəri]	*a*. 军事的
calculation [ˌkælkjuˈleiʃən]	*n*. 计算 e. g. Susan looked at the bill and made some rapid calculations. 苏珊看了一下账单，迅速地算了算。
encryption [inˈkripʃən]	*n*. 加密 e. g. Pay attention to secure encryption of data. 注意数据的安全加密。
decryption [diːˈkripʃən]	*n*. 解密

trajectory [trə'dʒektəri]	n. 轨道，轨线
crunch [krʌntʃ]	v. 压碎，扎，进行大量运算 e. g. The computer will crunch all the numbers to determine the final score. 计算机会计算所有的数字以得到最后的分数。
subsequent ['sʌbsikwənt]	a. 后来的
availability [əˌveilə'biləti]	n. 可用性，有效性
ubiquity [juː'bikwəti]	n. 普遍存在
accessible [ək'sesəbl]	a. 可存取的，可接近的，可进入的
breakthrough ['breikθruː]	n. 突破，突破性进展
syntax ['sintæks]	n. 语法，句法
crucial ['kruːʃəl]	a. 关键性的
modify ['mɔdifai]	v. 修改，变更 e. g. The regulations can only be modified by a special committee. 规则只能由一个特别委员会修改。
novice ['nɔvis]	n. 新手，初学者
grip [grip]	v. 紧握，支配 e. g. I gripped the rail and tried not to look down. 我抓住栏杆，尽量不往下看。
manipulation [məˌnipju'leiʃən]	n. 操作，处理

Key Terms

transistor 晶体管
integrated circuit 集成电路
programmer 程序员
chip 芯片
script 脚本
compiler 编译器

Useful Expressions

lead to 导致，通向

e. g. His carelessness led to the accident. 他的粗心大意导致了事故的发生。

by hand 用手

e. g. We had to wash our clothes by hand. 我们得手洗我们的衣服了。

deal with 处理，设计

e. g. Don't worry, I will deal with this. 别担心，我来处理。

result in 导致，结果是

be engaged in 从事于，忙于

e. g. He is engaged in design. 他从事设计工作。

Notes

1. Konrad Zuse

康拉德·祖萨，德国人，开发了世界上第一台采用二进制和布尔逻辑的可编程计算机。

2. Plankalkul

Zuse 在 1945 开发的世界上第一个编程语言。它惊人地完整，并在数据结构方面具有最先进的特性。它的某些特性与 40 年后的 ADA 语言非常相似。

3. Python

一种面向对象、直译式计算机程序设计语言，也是一种功能强大的通用型语言，已经具有二十多年的发展历史，成熟且稳定。

4. Visual Basic

可视化 Basic 语言。

5. PHP

一种 HTML 内嵌式语言，与微软的 ASP 颇有几分相似，都是一种在服务器端执行的嵌入 HTML 文档的脚本语言，语言的风格类似于 C 语言，现在被很多的网站编程人员广泛运用。PHP 独特的语法混合了 C、Java、Perl 以及 PHP 自创的新语法。

6. Perl

一种自由且功能强大的编程语言。自 1987 年初次登台亮相以来，它的用户数一直急剧膨胀。从最初被当作一种在跨平台环境中书写可移植工具的高级语言开始，Perl 就已经被广泛地视为一种工业级的强大工具，特别适合系统管理和 Web 编程。Perl 实际上已经被所有 UNIX（包括 Linux）捆绑在一起作为标准部件发布，而且也被广泛用于 Microsoft Windows 和几乎所有其他操作系统。Amiga、BeOS、VMS、MVS 和 Apple Macintosh 等也只是 Perl 已经完成移植的平台的一小部分。

7. FORTRAN

英文 formula translator 的缩写，译为“公式翻译器”。它是世界上最早出现的计算机高级程序设计语言，广泛应用于科学和工程计算领域。FORTRAN 语言以其特有的功能在数值、科学和工程计算领域发挥着重要作用。早在 1951 年，美国 IBM 公司的约翰·贝克斯（John Backus）就针对汇编语言的缺点着手研究开发 FORTRAN 语言，并于 1954 年在纽约正式对外发布。

Exercises

1. Match each of the following terms with its Chinese equivalent.

1) trajectory calculation — a. 商业软件
2) encryption — b. 微控制器
3) programmer — c. 机器代码
4) programming language — d. 操作系统
5) text manipulation — e. 程序员
6) scripting language — f. 加密
7) operating system — g. 弹道计算
8) comercial software — h. 程序语言
9) machine code — i. 文本处理
10) microcontroller — j. 脚本语言

2. Recognize the following abbreviations by matching them with their corresponding full names and translate them into Chinese.

1) PHP ________________ — a. object-oriented programming
2) Perl ________________ — b. application program interface
3) FORTRAN ________________ — c. algorithmic language
4) COBOL ________________ — d. programming in logic
5) LISP ________________ — e. simulation language
6) SIMULA ________________ — f. practical extraction and report language
7) PROLOG ________________ — g. list processing
8) ALGOL ________________ — h. common business-oriented language
9) API ________________ — i. formula translation
10) OOP ________________ — j. professional hypertext preprocessor

3. Complete the following sentences by translating the Chinese in the brackets.

1) Some military calculation needs were a driving force ________________ (在计算机发展初期).

2) ______________ (电子技术的发展) in the post-war years led to the construction of more practical electronic computers.

3) ______________ (程序语言的一个主要目标) is to enable programmers to express their intent for a computation more easily than they could with a lower-level language or machine code.

4) ______________ (程序语言是重要的工具) for helping software engineers write better programs faster.

5) ______________ (在过去的几十年间), a large number of computer languages have been introduced

4. Choose the best answer for each blank. (2009下半年程序员考试上午试题 (B))

1) The two goals of an operation system are ________ and efficiency.

A. convenience B. quick C. optimize D. standardize

2) A data ________ is a file that contains metadata—that is, data about data.

A. structure B. table C. base D. dictionary

3) Software ________ activities consume a large portion of the total life-cycle budget.

A. repair B. maintenance C. change D. update

4) A network software consists of ________, or rules by which processes can communicate.

A. protocols B. programs C. devices D. computers

5) The firewall device is a ________ system for connecting a computer network to other computer network.

A. hardware B. software C. security D. I/O

Supplementary Reading

Different Kinds of Programming Languages

Machine Language

Computer programs that can be run by a computer's operating system are called executables. An executable program is a sequence of extremely simple instructions known as machine code. These instructions are specific to the individual computer's CPU and associated hardware. For example, Intel Pentium and Power PC microprocessor chips each have different machine languages and require different sets of codes to perform the same task. Machine code instructions are few in number (roughly 20 to 200, depending on the computer and the CPU). Typical instructions are for copying data from a memory location or for adding the contents of two memory locations (usually registers in the CPU). Machine code instructions are binary—that is, sequences of bits (0 s and 1 s). Because these numbers are not understood easily by humans, computer instructions usually are not written in machine code.

Assembly Language

Assembly language is easier for programmers to understand than machine language. Each machine language instruction has an equivalent command in assembly language. For example, in assembly language, the statement " MOV A, B" instructs the computer to copy data from one location to another. The same instruction in machine code is a string of 16 0 s and 1 s. Once an assembly-language program is written, it is converted to a machine-language. It is still difficult to use, however, because assembly language instructions are a series of abstract codes. In addition, different CPUs use different machine languages and therefore require different assembly languages. Assembly language is sometimes inserted into a high-level language program to carry out specific hardware tasks or to speed up a high-level program.

High-level Languages

The improvement of machine language to assembly language set the stage for further advances. It was this improvement that led, in turn, to the development of high-level languages. If the computer could translate convenient symbols into basic operations, why couldn't it also perform other clerical coding functions.

A high-level programming language is a means of writing down, in formal terms, the steps that must be performed to process a given set of data in a uniquely defined way. It may bear no relation to any given computer but does assume that a computer is going to be used. The high-level languages are often oriented toward a particular class of processing problems. For example, a number of languages have been designed to process problems of a scientific-mathematic nature, and other languages have appeared that emphasize file processing applications.

Procedural Programming and Object-oriented Programming

There are two popular approaches to writing computer programs: procedural programming and object-oriented programming.

Procedural programming involves using your knowledge of a programming language to create computer memory locations that can hold values and writing a series of steps or operations that manipulate those values. For convenience, the individual operations used in a computer program are often grouped into logical units called procedures. A procedural program defines the variable memory locations and then calls or invokes a series of procedures to input, manipulate, and output the values stored in those locations. A single procedural program like C language program often contains hundreds of variable and thousands of procedure calls.

Object-oriented programming is an extension of procedural programming in which you take a slightly different approach to writing computer programs. Thinking in an object-oriented manner involves envisioning program components as objects that are similar to concrete objects in the real world. Then you manipulate the objects to achieve a desired result. Writing object-oriented programs involves both creating objects and creating applications that use those objects.

Object-oriented programming (OOP) languages like C++ are based on traditional high-level languages, but they enable a programmer to think in terms of collections of cooperating objects instead of lists of commands.

New Words and Expressions

sequence ['si:kwəns] *n*. 顺序，序列
binary ['bainəri] *a*. 二进制的，二元的
equivalent [i'kwivələnt] *a*. 等价的，相等的
string [striŋ] *n*. 一串 *v*. 串起
insert [in'sə:t] *vt*. 插入
uniquely [ju'ni:kli] *ad*. 独特地
approach [ə'prəutʃ] *n*. 方法，途径 *v*. 靠近，着手处理
manipulate [mə'nipjuleit] *vt*. 操纵，操作
procedure [prə'si:dʒə(r)] *n*. 程序，步骤
invoke [in'vəuk] *vt*. 调用
extension [ik'stenʃən] *n*. 延长
envisioning [en'viʒəniŋ] *n*. 想象，射向
Intel Pentium 英特尔奔腾处理器
assembly language 汇编语言
speed up 加速
procedural programming 程式编程
object-oriented programming 面向对象程序设计
memory location 存储单元
logical unit 逻辑单元
procedure calls 过程调用，程序呼叫

Reading Comprehension

Read the following statements below and decide if they are true (T) or false (F) according to the passage you have just read.

1) Intel Pentium and Power PC microprocessor chips have common machine language and require the same sets of codes to perform the same task. (   )
2) Compared with machine language, assembly language is easier for programmers to understand. (   )
3) It was the improvement of machine language that led, in turn, to the development of high-level languages. (   )
4) Writing object-oriented programs involves only creating objects. (   )

Part 3 Screen English

电源管理设置

项　目	参　数
Power Management　电源管理	Disabled/User Define/Min Saving/Max Saving　关闭电源管理功能/用户自行定义/使用省电量最小的模式/使用省电量最多的模式
Video Off Option　显示器关闭选项	Suspend Off　挂起状态显示器关闭 All Mode Off　所有模式显示器关闭 Always On　所有模式显示器都不关闭
HDD Power Down　硬盘停止工作	Disabled/1～15min　不设置硬盘停止工作时间/设定在1～15分钟后硬盘停止工作
Date（of Month）　系统开机日期（按月计）	0～31　0：系统在任何一天都可开机；1～31：系统开机日期
Timer（hh：mm：ss）　系统开机时间（小时：分：秒）	00：00：00～23：59：59
HDD Down in Suspend　系统挂起时硬盘不工作	Enabled/Disabled　系统挂起时硬盘不工作/系统挂起时硬盘正常工作

提示

在系统开机提示信息及CMOS参数设置中，大量出现名词短语及用缩略形式表示的专业术语，而非采用完整的英语句子的形式，这在计算机屏幕英语的阅读中，须重点加以留意。

Quotation

There's an old story about the person who wished his computer were as easy to use as his telephone. That wish has come true，since I no longer know how to use my telephone. —— Bjarne Stroustrup

以前有这么一个故事，有一个人希望计算机像电话一样容易使用。这个愿望后来实现了，因为我也不知道怎样使用我的电话了。

Key to Exercises

Listening & Speaking

1. 1) instructions　　2) machine　　3) programs　　4) invention of the computer

5) imperative　　6) functional programming　　7) logic programming

2. 1) B　2) B　3) A　4) A　5) A

Exercises

1. 1) g　2) f　3) e　4) h　5) i　6) j　7) d　8) a　9) c　10) b
2. 1) j 超文本预处理器（语言）　2) f 实用提取与报告语言　3) i 公式翻译器（语言）
   4) h 面向商业的通用语言　5) g 表处理（语言）　6) e 仿真语言
   7) d 逻辑程序设计语言　8) c 算法语言　9) b 应用程序编程接口
   10) a 面向对象程序设计
3. 1) in early computer development
   2) The advances in electronic technology
   3) A primary purpose of programming languages
   4) Programming languages are important tools
   5) During the last few decades
4. 1) A　2) D　3) B　4) A　5) C

Comprehension

1) F　2) T　3) T　4) F

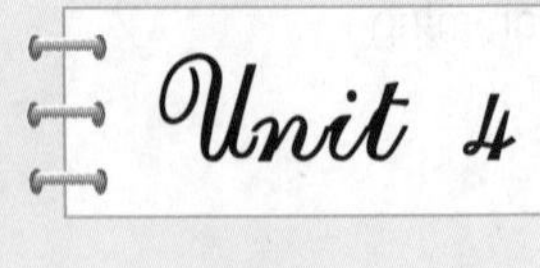

Office Automation

Learning Objectives

After completing this unit, you will be able to:

1. Know e-mail and BBS;
2. Learn something about office automation system.

Fax Machine

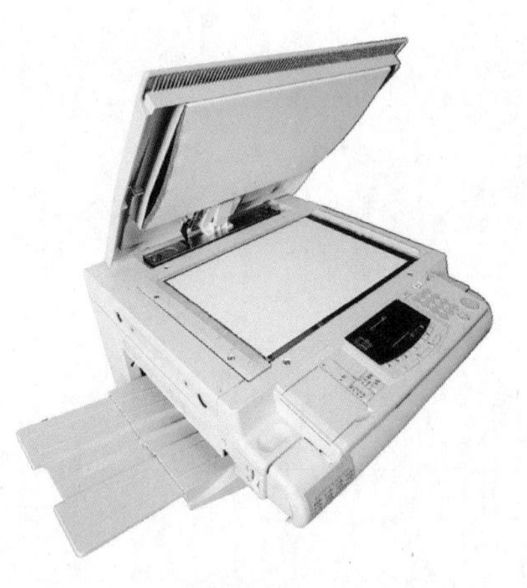

Copier

Part 1 Listening & Speaking

1. Listen to the following passage and fill in the blanks with the words in the box.

displaying	messages	effort	received	network	users

Electronic Mail

Millions of end 1) ________ now depend on electronic mail (e-mail) to send and receive electronic 2) ________. You can send e-mail to anyone on your 3) ________ for storage in his/her electronic mail boxes or magnetic disk drives. Whenever they are ready, they can read their electronic mail by 4) ________ it on the video screens at their workstations. So, with only a few minutes of 5) ________ (and a microseconds of transmission), a message to one or many individuals can be composed, sent, and 6) ________.

2. Choose the proper words or expressions.

Bulletin Board Systems (BBS)

Bulletin Board Systems are a 1) ________ telecommunications services provided by the Internet, public information services, and 2) ________ of business firms, organization, and end user groups. An electronic bulletin board system allows you to post public or private messages that other end 3) ________ can read by accessing the BBS with their computers. Establishing a small BBS for a business is not that difficult. Minimum requirements are a microcomputer with a hard disk drive, custom or packaged BBS 4) ________, modem and a telephone line. Bulletin board system serves as a 5) ________ location to post and pick up messages or upload and download data files or programs 24 hours a day. A BBS helps end users ask questions, get advice, locate and share information, and get 6) ________ touch with other end users.

1) A. popular　　B. unpopular

2) A. thousand　　B. thousands

3) A. use　　B. users

4) A. software　　B. hardware

5) A. centre　　B. central

6) A. on　　B. in

3. Please read the conversation below and learn how to raise a question and how to solve each problem.

秘书 secretary	专业 major	设备 equipment
文件 document	打字 type	简历 resume
盼望 look forward to		

A：早上好！请问你们招人吗？

B：早上好！我们缺一位秘书。

A：我正好是秘书专业的。能让我试试吗？

B：当然。办公室里面的设备你都会使用吗？

A：我会使用计算机、传真机和复印机。

B：好的。秘书要处理很多文件，我想知道你打字的速度有多快？

A：我一分钟能打 60 个字。

B：好的。请把你的简历发到这个邮箱 Sally123@yahoo. com. cn。星期五我们会给你打电话。

A：邮箱地址是 Sally123@yahoo. com. cn，对吗？

B：是的。

A：谢谢您。盼望您的来电。

B：不客气，再见。

A：再见。

A：Good morning！Is there any job vacancy?

B：Good morning！We have a vacancy for a secretary.

A：That's just my major. Can I have a try?

B：Of course. Do you know how to use the equipment in an office?

A：Yes，I can use computers，fax machines and digital copiers.

B：Good. And as you know，a secretary will have to deal with many documents，so I'd like to know how fast you can type?

A：I can type 60 words per minute.

B：Ok，please send your resume to this email address：Sally123@yahoo. com. cn. We will give you a call on Friday.

A：The address is Sally123@yahoo. com. cn，right?

B：Yes，it is.

A：OK，thank you for your information and time. I am looking forward to your call.

B：You are welcome. Goodbye.

A：Goodbye.

4. Oral Practice

Sally 想要给 Linda 通过电子邮件发一份生日贺卡，但是她不知道该怎么做，于是向

John 请教。

Key Words

登陆 access	账户 account	收件人 receiver
主题 subject	单击 click	附件 attachment

Part 2 Reading

Text

Office Automation

Office automation has been viewed as telecommunication-based information systems that collect, process, store, and distribute electronic messages, documents and other forms of communications among individual work groups and organizations. Such systems can improve the collaboration and productivity of secretarial staff and work groups by significantly reducing the time and efforts needed to produce, distribute, and share office information.

Q1 What does office automation mean?

The entire office automation system comprises the following five sub-systems: electronic publishing system, electronic communication system, electronic collaboration system, image processing system, and office management system.

Q2 What are the five sub-systems of office automation?

Electronic Publishing System

Electronic publishing system has transformed today's office into an in-house publisher of office documents. Word processing is the use of computer systems to create, edit, revise and print text materials. With desktop publishing, organizations can design and print their own newsletter, brochures, manuals and books with several type styles, graphics and colors on each page.

Q3 What has electronic publishing system transformed today's office into?

Q4 What is word processing?

Q5 What can desktop publishing do?

Electronic Communication System

Electronic mail (e-mail), voice mail, bulletin board systems, and facsimile allow organizations to send messages in text, video or voice form or transmit copies of document and to receive it in seconds.

Q6 Can organizations send messages in video and voice?

Electronic Collaboration System

Electronic meeting system involves the use of video and audio communications to allow conference and meeting to be held with participants who may be scattered across a room, a building, a country, or the globe. Reducing the need to travel to and from meetings should save employee time, increase productivity, and reduce travel expenses and energy consumption.

Q7 With the help of electronic meeting system, do people still have to take a plane to have meetings?

Image Processing System

Image processing system allows end users to electronically capture, store, process, and retrieve images of documents that may include numeric data, text, handwriting, graphics, and photographs.

Q8 What can image processing system do?

Office Management System

Office management system provides computer-based support services to managers and other professionals to help them organize their work activities. Office management software computerizes manual methods of planning such as paper calendars, appointment books, directories, file folders, memos, and notes.

Q9 What can office management system do?

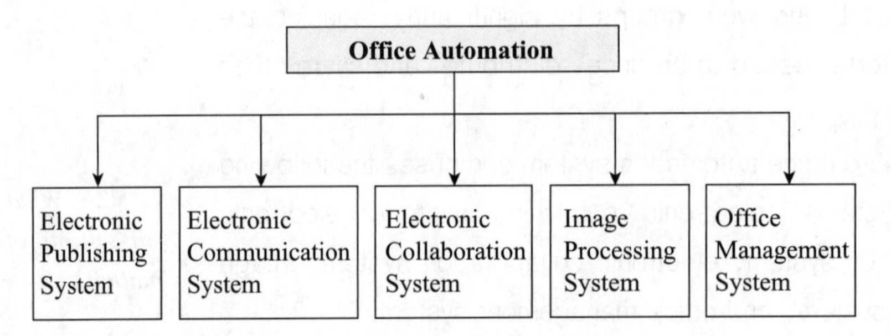

An Overview of An Office Automation System

New Words

distribute [di'stribjuːt]	*vt*. 分配，散布，分开，把……分类 e. g. They distributed food among the poor. 他们为穷人分发食物。
collaboration [kəˌlæbə'reiʃən]	*n*. 合作，勾结 e. g. Microsoft also fixed a flaw in Groove, which is an Office program that facilitates document sharing and collaboration. 微软修复了Groove中的一个漏洞。Groove是一个促进文件共享和合作的Office程序。

significantly [sig'nifəkəntli]	*ad*. 意味深长地，意义重大地
comprise [kəm'praiz]	*vt*. 包含，由……组成
transform [træns'fɔːm]	*v*. 改变，使……变形，转换 e. g. They transformed the basement into a reading room. 他们把地下室改作阅览室。 to transform dream into reality 变梦想为现实
in-house ['inˌhaus]	*a*. 内部的　*ad*. 内部地
desktop ['desktɔp]	*n*. 桌面，台式计算机 e. g. Notebook sales have overtaken those of desktop PCs in the consumer market. 在消费者市场上，便携式计算机的销量已经超过台式机。
newsletter ['njuːzˌletə]	*n*. 时事通讯
brochure [bro'ʃur]	*n*. 手册，小册子
manual ['mænjuəl]	*a*. 手工的，体力的　*n*. 手册，指南
facsimile [fæk'simili]	*n*. 传真，复写　*a*. 复制的 e. g. As a technology，facsimile transmission is more than a century old. 作为一项技术，传真的历史已有一个多世纪了。
video ['vidiəu]	*n*. 视频，录像，录像机，电视
transmit [trænz'mit]	*v*. 传输，传播，发射，传达，遗传 e. g. The program's main feature is the cartoon character a user can transmit along with his or her e-mail. 这种程序的主要特点是在用户发送电子邮件时可以同时发送一个卡通人物。
audio ['ɔːdiəu]	*a*. 声音的，音频的，声频的
scatter ['skætə]	*v*. 分散，散开，散射
consumption [kən'sʌmpʃən]	*n*. 消费，消耗

retrieve [ri'triːv]	*v*. 检索，恢复，重新得到 e. g. The display has a memory card interface that can store and retrieve pre-recorded data such as maps, reconnaissance information and mission plans. 这种显示器配有存储卡接口，能够储存并检索预先录制的数据，如地图、侦察信息和任务规划等。
numeric [njuː'merik]	*a*. 数值的（=numerical）

Key Terms

electronic publishing system 电子出版系统

electronic communication system 电子通信系统

electronic collaboration system 电子协同系统

image processing system 影像处理系统

office management system 办公管理系统

word processing 文字处理（系统），文字信息处理

desktop publishing 桌面出版，桌面排版印刷，台式刊印，台式印刷

voice mail 语音信箱，电话录音传送系统，语音邮件

end user 终端用户

Useful Expressions

scatter across 散布，分散

e. g. scatter across the world 散布于世界各地

be scattered across Thailand 分散在泰国各地

Exercises

1. Match each of the following terms with its Chinese equivalent.

1) distribute	a. 传输
2) collaboration	b. 包含
3) comprise	c. 检索
4) in-house	d. 传真
5) brochure	e. 分配
6) facsimile	f. 数字的
7) transmit	g. 合作
8) audio	h. 内部的
9) retrieve	i. 手册
10) numeric	j. 音频

2. Recognize the following abbreviations by matching them with their corresponding full names and translate them into Chinese.

1) EPS ________	a. office management system
2) ECS ________	b. electronic collaboration system
3) ECS ________	c. image processing system
4) IPS ________	d. electronic publishing system
5) OMS ________	e. electronic communication system
6) HPFS ________	f. office automation
7) LFN ________	g. uninterruptable power supply
8) USB ________	h. universal serial bus
9) UPS ________	i. high performance file system
10) OA ________	j. long file name

3. Complete the following sentences by translating the Chinese in the brackets.

1) Office automation has been viewed as telecommunication-based information systems that ________ (收集、处理、存储、分配) electronic messages, documents and other forms of communications among individual work groups and organizations.

2) The entire ________ (办公自动化系统) comprises five sub-systems.

3) Word Processing is the use of computer systems to ________ (创建、编辑、修改和打印) text materials.

4) Electronic meeting systems involve the use of ________ (视频和音频) communications

5) Image Processing System allows ________ (终端用户) to electronically capture, store, process, and retrieve images of documents

4. Choose the best answer for each blank. (2009 年下半年程序员考试上午试题 (B))

1) We can use the word processor to ________ your documents.

A. edit B. compute C. translate D. unload

2) ________ software, also called end-user program, includes database programs, word processors, spreadsheets etc.

A. Application B. System C. Compiler D. Utility

3) ________ processing offers many ways to edit text and establish document formats. You can easily insert, delete, change, move and copy words or blocks of text.

A. Data B. Database C. Word D. File

4) A ________ copies a photograph, drawing or page of text into the computer.

A. scanner B. printer C. display D. keyboard

5) ________ is the sending and receiving of the message by computer. It is a fast, low-

cost way of communicating worldwide.

A. LAN B. Post office C. E-mail D. Interface

Supplementary Reading

How to Write in Word?

You need to create a document in Word, but you've never worked with Word. Where do you begin? Or perhaps you've worked in Word a time or two, but you still wonder how to do some of the basics—edit and format text, or change margins. If you need to learn the skills to get to work in Word quickly, with little fuss, this is the place.

When you start Word, a new file opens. That file is called a document. Above the document you will see the menu bar and the tool bars displayed across the top of the window. Click anywhere of the document, in the upper-left corner of the document, or page, there will be the insertion point, a blinking vertical line. The first thing you type will appear there. You can start further down the page if you want to by pressing ENTER until the insertion point is where you want the first line to begin. As you type, the insertion point moves to the right.

If you're typing a letter, you might start by typing the date. After that, press ENTER to move the insertion point down the page by one line. If there's an address, you might add a few empty lines before you type. Press ENTER several times until the insertion point is where you want it. Then type the greeting. When you type the body of the letter, if you want to indent the first line of a paragraph, you can do that by pressing the TAB key on your keyboard to move the insertion point one-half inch to the right. Go ahead and type. When you get to the end of the first line, you don't have to press ENTER as you would. Word takes care of that for you. Just continue to type. Whatever you are typing will continue on to the next line. You do press ENTER to start another paragraph.

New Words and Expressions

margin [ˈmɑːdʒin] *n*. 页边的空白，边缘，利润，盈余

fuss [fʌs] *n*. 奔忙，瞎忙，小题大做

insertion [inˈsəːʃən] *n*. 插入，嵌入，插入物

blinking [ˈbliŋkiŋ] *a*. 闪烁的

menu bar 菜单栏

tool bar 工具栏

TAB key 跳格键，制表键

Reading Comprehension

Read the following statements below and decide if they are true (T) or false (F) according to the passage you have just read.

1) We can use Word to edit and format text, or change margins. ( )

2) At the bottom of the document you will see the menu bar and the tool bars. (   )

3) If you want to start further down the page, you shall press Backspace. (   )

4) If you want to indent the first line of a paragraph, you can do that by pressing the TAB key on your keyboard. (   )

5) When you get to the end of the first line, you must press ENTER to come to the next line. (   )

Part 3 Screen English

电子邮件的常用操作

提示信息	含 义
(1) To open a message Click the underlined text in the From column of a message you wish to open. There will be a red arrow to the left of any message that you haven't opened yet.	(1) 打开信息 单击 From 列中你想打开信息的加有下划线的文本。在你尚未打开信息的左边将有一个红箭头予以提示。
(2) To save a message You can save messages as follows: 1) Open the message you want to save. 2) On the menu bar of your browser, click File. 3) On the File[①] menu, click Save As. Indicate where you want to save the message on your computer. Give the message a file name that will help you locate it. Indicate what type of file you want to save it as. Click Save. If you decide you don't want to save the file, click Cancel. You can also save a copy of a message by using Outlook Express to move it to a local folder in Outlook Express.	(2) 保存信息 你可按如下步骤保存信息: 1) 打开你想要保存的信息。 2) 在浏览器的菜单栏上，单击“文件”。 3) 在“文件”菜单上，单击“另存为”。 指明该信息将被存放在计算机上的位置。给信息取一个文件名以帮助你寻找该信息。指明想要保存的文件类型。单击“保存”。 如果你不想保存文件，单击“取消”。 你也可以使用 Outlook Express 把该信息移入 Outlook Express 的本地文件夹中，作为信息的备份予以保存。

（续）

提示信息	含　义
(3) To send a file with your message 1) Click Compose on the horizontal navigation bar. 2) On the Compose page, click Attachments. 3) On the Attachments page, specify which file(s) you want to send with your message. For each file you specify, click Attach to Message. When you have added all the files you wish to, click Done. You'll return to the Compose page. Complete and send your message. You can attach more than one file, but the total size of your message including attachments can't be larger than 1 megabyte (MB) and no one attached file can be larger than 500 kilobytes (KB).	(3) 将信息与文件一起发送 1) 单击水平导航栏上的“写信”。 2) 在“写信”页面中，单击“附件”。 3) 在“附件”页面中，指定要与信息一起发送的文件。 对于每个所指定的文件，单击“附于信息”。添加需要的所有文件后，单击“完成”。 返回“写信”页面。完成并发送信息。你可在信息上添加一个以上的附件，但连附件在内信息的总容量不能超过1兆字节（MB），并且每个附件的大小不超过500千字节（KB）。
(4) To print a message Open the message you want to print. On the menu bar of your browser, click File. On the File menu, click Print. Complete the items in the Print dialog box as you would to print anything else. Click Print. If you decide you don't want to print the message, click Cancel. If you're not using Internet Explorer 4.0 or later or Netscape Navigator 4.0 or later, you can print a message by copying its content and pasting the text into a word-processing program such as Word.	(4) 打印信息 打开想要打印的信息。在浏览器菜单栏上，单击“文件”。在“文件”菜单上，单击“打印”。 正如你要打印其他内容一样，完成“打印”对话框中的选项设置，然后单击“打印”。 如果不打印信息，单击“取消”。 如果使用的浏览器不是Internet Explorer 4.0或更高版本、Netscape Navigator 4.0或更高版本，你可以通过将信息内容复制并粘贴到如Word这类文字处理程序中来打印信息。

注①：“On the File menu, click Save As.”中的“File”“Save As”均为操作命令或按钮的名称。

Quotation

Before you were born, your parents weren't as boring as they are now. They got that way from paying your bills, cleaning your clothes and listening to you talk about how cool you are. So before you save the rain forest from the parasites of your parents' generation, try "delousing" the closet in your own room. ——Bill Gates

在你出生之前，你的父母并非像他们现在这样乏味。他们变成今天这个样子是因为这些年来他们一直在为你付账单，给你洗衣服，听你大谈你是如何的酷。所以，如果你想消灭你父母那一辈中的"寄生虫"来拯救雨林的话，还是先试着给你房间衣柜里"灭虱"吧。

If you think your teacher is tough, wait till you get a boss. He doesn't have tenure.

——Bill Gates

如果你认为你的老师严厉，等你有了老板就不会这样想了。老板可是没有任期限制的。

Key to Exercises

Listening & Speaking

1. 1) users 2) messages 3) network 4) displaying 5) effort 6) received
2. 1) A 2) B 3) B 4) A 5) B 6) B

Exercises

1. 1) e 2) g 3) b 4) h 5) i 6) d 7) a 8) j 9) c 10) f
2. 1) d 电子出版系统 2) b 电子协同系统 3) e 电子通信系统
   4) c 影像处理系统 5) a 办公管理系统 6) i 高性能文件系统
   7) j 长文件名 8) h 通用串行总线 9) g 不间断电源
   10) f 办公自动化
3. 1) collect, process, store and distribute
   2) office automation system
   3) create, edit, revise and print
   4) video and audio
   5) end users
4. 1) A 2) A 3) C 4) A 5) C

Comprehension

1) T 2) F 3) F 4) T 5) F

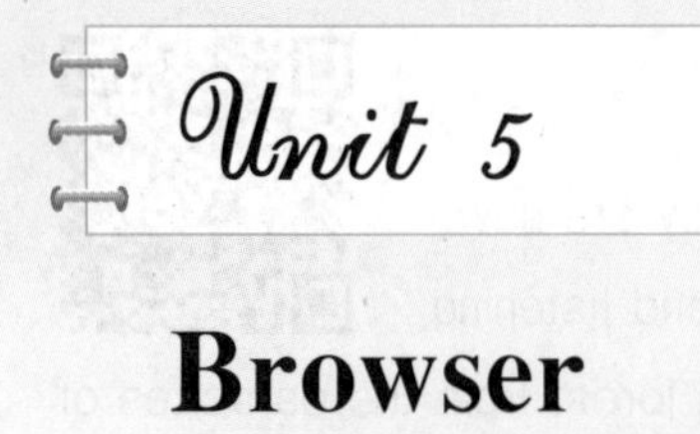

Unit 5 Browser

Learning Objectives

After completing this unit, you will be able to:

1. Know the definition of browser;
2. Explain the features of browser;
3. Use a browser on one of the platforms.

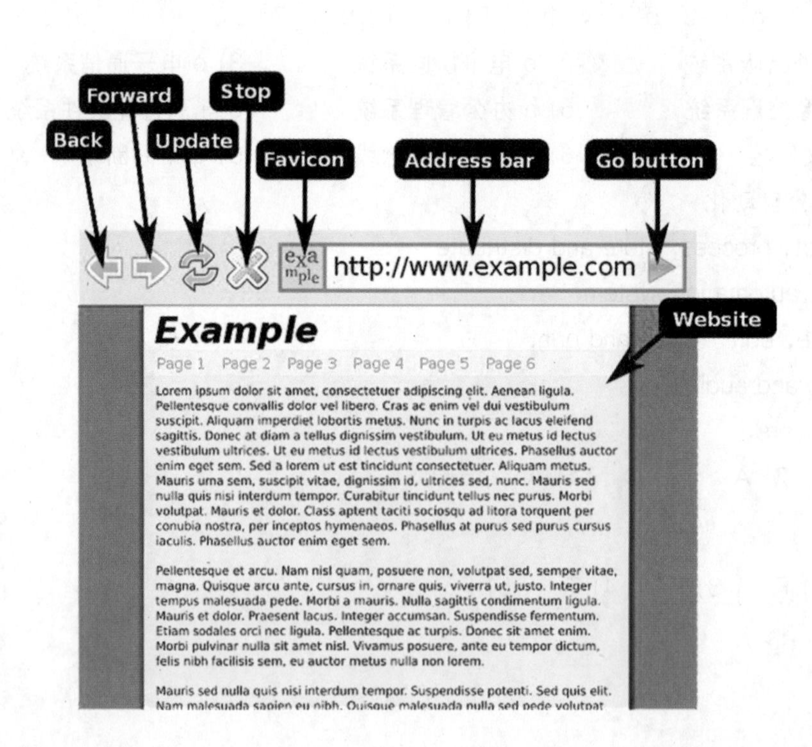

Part 1 Listening & Speaking

1. Listen to the following passage and fill in the blanks with the words in the box.

hacker	holes	apply	install	addresses	released
manually	resolved	component	font		

Microsoft Patches IE 9 with New Security Update

Microsoft has 1) ________ a new update for Internet Explorer 9 that aims to patch several outstanding security 2) ________.

Available through Windows Update since October 13, 2011, the security update is rated critical by Microsoft, which means that people who have Windows Update set to "3) ________ updates automatically" will automatically receive it.

Users who haven't enabled that option are advised to install the update 4) ________ from Windows Update. IT administrators who support large organizations should also 5) ________ the update with whatever patch management software they use in-house.

The update targets eight vulnerabilities in IE 9, some of which could let a 6) ________ remotely run code on a PC if the user visits a "specially crafted Web page" using Microsoft's browser. Such an exploit could allow someone to gain the same rights on the PC as the local user. The update specifically changes the way IE allocates and 7) ________ memory, according to Microsoft's Security Bulletin.

Moving Internet Explorer 9 up to version 9.0.3, the update also fixes holes in earlier versions of IE, specifically 6, 7, and 8. A number of non-security related issues are 8) ________ as well, including one that prevented users from changing the 9) ________ size in Windows Mail after installing IE 9 and another in which Windows 7 gadgets may not have worked properly.

This latest fix for IE is one 10) ________ of a larger security update that Microsoft rolled out this week as part of its monthly patch program. Beyond resolving holes in the browser, the series of patches addressed vulnerabilities in Windows, Silverlight, and the .NET Framework.

Ironically, the IE security fix comes at the same time that Microsoft has rolled out a new Web page that compares the security of the major browsers and found IE to be the most secure of them all.

2. Choose the proper words or expressions.

What Are Browser Helper Objects?

From this point of view, Internet Explorer is just like any other Win32-based program with its own memory space to preserve. With Browser Helper Objects you can write components—specifically, in-process Component Object Model (COM) components—that Internet Explorer will 1) ________ each time it starts up. Such objects run in the same memory context as the browser and can perform any action on the available windows and modules. For example, a BHO could detect the browser's typical events, such as GoBack, GoForward, and DocumentComplete; access the browser's menu and 2) ________ and make changes; create windows to display additional information on the currently viewed page; and 3) ________ hooks to monitor messages and actions.

Before going any further with the nitty-gritty details of BHO, there are a couple of points needed to illuminate further. First, the BHO is 4) ________ to the browser's main window. In practice, this means a new instance of the object is created as soon as a new browser window is created. Any instance of the BHO lives and dies with the browser's instance. Second, BHOs only exist in Internet Explorer, version 4.0 and later.

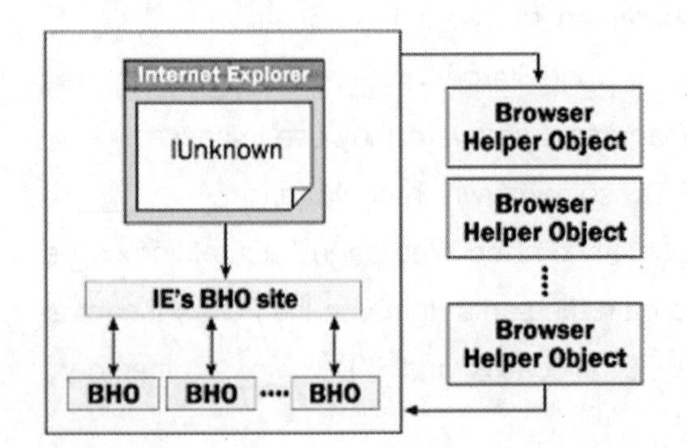

If you're running the Microsoft Windows 98, Windows 2000, Windows 95, or Windows NT version 4.0 operating system with the Active Desktop Shell Update (shell version 4.71), BHOs are supported also by Windows Explorer. This has some implications that will be talked more about later when making performance considerations and 5) ________ the impact of BHOs.

In its simplest form, a BHO is a COM in-process 6) ________ registered under a certain registry's key. Upon startup, Internet Explorer looks up that key and loads all the objects whose CLSID is stored there. The browser initializes the object and asks it for a certain interface. If that interface is found, Internet Explorer uses the methods provided to pass its unknown pointer down to the helper object.

1) A. lock　　　B. load

2) A. toolbox　　　B. toolbar

3) A. install　　　B. instate

4) A. tied B. tired
5) A. value B. evaluating
6) A. server B. serve

3. Please read the conversation below and learn how to raise a question and how to solve each problem.

下载 download	安装 install	运行 run
系统 system	站点 site	定位 locate
双击 double-click	指示 instruction	更新程序 updating program
网络连接 network connection		手动 manually

A：有什么需要帮忙吗?
B：我想下载和安装 IE 9，你能不能告诉我该怎么做呢?
A：首先，为了能安装和运行 IE 9，你必须有 Windows 7 或者 Windows Server 2008 系统。Windows XP 系统是不支持 IE 9 的.
B：哦，我的电脑系统是 Windows 7。
A：那就好。然后你可以从一些站点下载 IE 9。一旦完成了名为 IE9setup. exe 文件的下载，直接定位到该文件，接着双击，然后按照指示做就可以了。
B：如果安装时提示更新程序安装失败，我该怎么办呢?
A：这可能是由于网络连接问题导致的。
B：那这时我该怎么办呢?
A：你可以手动下载安装这些更新程序，问题即可解决。

A：What can I do for you?
B：I'd like to download and install IE 9. Could you tell me how to do?
A：First, in order to install and run IE 9, you must have Windows 7 or Windows Server 2008 system. Windows XP system does not support IE 9.
B：Oh, my computer system is Windows 7.
A：It's great. You can begin your IE 9 download at some sites. Once you finish downloading the setup file, which is named IE9setup. exe, simply locate this executable file, double-click on it, and follow the instructions.
B：What can I do if I'm prompted that the installation of the updating program fails?
A：It may be caused by the network connection.
B：Then what should I do?
A：You can download and install the updating program manually to solve this problem.

4. Oral Practice

Jim 找不到 IE 9 浏览器中的收藏夹按钮了。请你告诉他在 IE 9 中，收藏夹按钮从左边移至右边了，并且与原来 IE 8 不同，新按钮只有一个星号图标，旁边不再显示“收藏夹”字样。

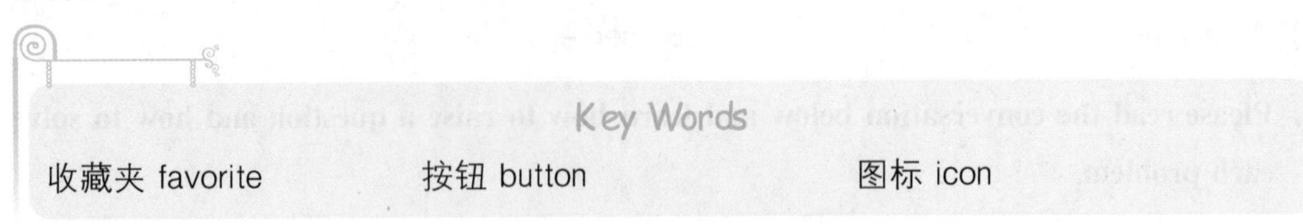

Key Words

收藏夹 favorite　　按钮 button　　图标 icon

Part 2 Reading

Text

Browser

Definition

A Web browser or Internet browser is a software application for retrieving, presenting, and traversing information resources on the World Wide Web (WWW). An information resource is identified by a uniform resource identifier (URI) and may be a Web page, image, video, or other piece of content. Hyperlinks present in resources enable users to easily navigate their browsers to related resources.

Although browsers are primarily intended to access the WWW, they can also be used to access information provided by Web servers in private networks or files in file systems. Some browsers can also be used to save information resources to file systems.

Function

The primary purpose of a Web browser is to bring information resources to the user. This process begins when the user inputs a URI, for example http://en.wikipedia.org/, into the browser. The prefix of the URI determines how the URI will be interpreted. The most commonly used kind of URI starts with "http:" and identifies a resource to be retrieved over the hypertext transfer protocol (HTTP). Many browsers also support a variety of other prefixes, such as

Q1 What is a browser?

Q2 What kind of browser do you use now?

Q3 What can we do with a browser?

Q4 Briefly describe the functions of a browser.

Q5 List the types of the file that most browsers can display.

"ftp:" for the file transfer protocol (FTP), and "file:" for local files.

In the case of http, https, file, and others, once the resource has been retrieved the Web browser will display it. HTML is passed to the browser's layout engine to be transformed from markup to an interactive document. Aside from HTML, Web browsers can generally display any kind of content that can be part of a Web page.[1] Most browsers can display images, audio, video, and XML files, and often have plug-ins to support Flash applications and Java applets.

Features

Available Web browsers range in features from minimal, text-based user interfaces with bare-bones support for HTML to rich user interfaces supporting a wide variety of file formats and protocols. Browsers which include additional components to support e-mail, Usenet news, and Internet relay chat (IRC), are sometimes referred to as "Internet suites" rather than merely "Web browsers".[2] Most Web browsers can display a list of Web pages that the user has bookmarked so that the user can quickly return to them. Bookmarks are also called "Favorites" in Internet Explorer.

Q6 What is the meaning of Internet suites?

Q7 What does "Favorites" mean in Internet Explorer?

User Interface

Most major Web browsers have these user interface elements in common:

- *Back*[3] and *Forward*[4] buttons to go back to the previous resource and forward again;
- A *History*[5] list, showing resources previously visited in a list (typically, the list is not visible all the time and has to be summoned);
- A *Refresh*[6] or *Reload*[6] button to reload the current resource;
- A *Stop*[7] button to cancel loading the resource. In some browsers, the Stop button is merged with the Reload button;

Q8 If you want to go back to where you started, click the ________.

Q9 If you want to return to the previous page, click the ________.

- A *Home*[8] button to return to the user's home page;
- An address bar to input the URI of the desired resource and display it;
- A search bar to input terms into a search engine;
- A status bar to display progress in loading the resource and also the URI of links when the cursor hovers over them, and page zooming capability.

New Words

application [ˌæpliˈkeiʃən]	*n*.［计算机］应用，应用程序（软件，包括文字处理程序、图形处理程序和数据表格程序等） e. g. No application is associated with the specified file name extension. 没有与指定文件扩展名相关联的应用程序。
retrieve [riˈtriːv]	*vt*.［美］［计算机］检索 e. g. They often retrieve some data from a disk. 他们常常从存储磁盘中检索一些信息。
traverse [trəˈvəːs]	*vt*. 仔细检查，详细讨论，全面研究 e. g. We need to traverse the last century thoroughly. 我们需要全面研究上一个世纪。
identify [aiˈdentifai]	*vt*. 辨认，识别，认出，鉴定，验明，确定 e. g. He identified his baggage among hundreds of others. 他在几百件行李中认出了自己的行李。
video [ˈvidiəu]	*a*. 录像的，视频的 e. g. In the medium term, Tudou wants to transform itself from a video distribution platform into a full media company. 土豆网的中期目标是希望从一个视频发布平台转型为一家全面的媒体公司。
navigate [ˈnævigeit]	*vt*. 驾驶，操纵，查明并指出……的航向（或方位），为（船只、飞机）导航，指引，指导 e. g. Applications or Web sites with dozens of distinct types of pages, screens, or forms are difficult to navigate. 如果应用程序或网站有几十个不同类型的页面、屏幕和表格，则通常难以导航。

access ['ækses]	*vt.* [计算机] 取得（数据），访问，存取（泛指取数或存数），在……上存取数据，取数，进网，接通（计算机），（信息）输入存储装置 e. g. Branch officials can access the central database. 分行官员可访问中央数据库。
prefix ['pri:fiks]	*n.* 前缀，词头 e. g. These network entries have a prefix length associated with them. 这些网络条目有一个与它们关联的前缀长度。
via ['vaiə]	*prep.* 经由，经过，取道，通过，凭借 e. g. The news reached us via a friend of mine. 这个消息是通过我的一个朋友传到我们这里的。
cursor ['kə:sə]	*n.* [计算机]（显示器的）光标 e. g. The cursor is the visible representation of the mouse's position on the screen. 光标是屏幕上鼠标位置的可视代表物。
zoom [zu:m]	*v.* 影像放大，增速，图像电子放大，变焦距 e. g. You can browse the entire page and then zoom in the content you need. 你可以浏览整个网页，然后放大所需要的内容。

Key Terms

uniform resource identifier (URI) 统一资源标识符
Web page 网页
hyperlink 超链接，超级链接
World Wide Web (WWW) 万维网
Web server 网络服务器
private network 专用网络
file system 文件系统
hypertext transfer protocol (HTTP) 超文本传输协议
file transfer protocol (FTP) 文件传输协议
hypertext markup language (HTML) 超文本标记语言
user network (Usenet) 用户网
Internet relay chat (IRC) 因特网中继聊天（程序）
plug-in 插件程序，插件法
user interface 用户界面，用户接口

Useful Expressions

intend to 打算做……，想要……

e. g. Where do you intend to spend your summer vacation this year? 今年你打算在什么地方度过暑假？

range from... to...（在……范围内）变动，变化

e. g. The prices of leather shoes range from 100 to 250 dollars. 皮鞋的价格从 100 美元到 250 美元不等。

merge with 与……结合，与……联合，融入（车流中）

e. g. The 20th century's revolution in information technology will thereby merge with the 21st century's revolution in biotechnology. 20 世纪信息技术的革命将由此同 21 世纪的生物技术革命融为一体。

Notes

1. Aside from HTML, Web browsers can generally display any kind of content that can be part of a Web page.

that 引导的定语从句修饰 content，that 在从句中作主语。

2. Browsers which include additional components to support e-mail, Usenet news, and Internet relay chat (IRC), are sometimes referred to as "Internet suites" rather than merely "Web browsers".

which 引导的定语从句修饰主语 Browsers。

3. Back　后退

4. Forward　前进

5. History　历史

6. Refresh/Reload　刷新

7. Stop　停止

8. Home　主页

Exercises

1. Match each of the following terms with its Chinese equivalent.

1) software application	a. 插件程序
2) Web page	b. 状态栏
3) hyperlink	c. 搜索引擎
4) file system	d. 软件应用
5) plug-in	e. 超级链接
6) search engine	f. 文件系统
7) address bar	g. 网页
8) status bar	h. 地址栏

2. Recognize the following abbreviations by matching them with their corresponding full names and translate them into Chinese.

1）WWW ________________	a. file transfer protocol
2）URI ________________	b. World Wide Web
3）HTTP ________________	c. Internet relay chat
4）FTP ________________	d. uniform resource identifier
5）IRC ________________	e. hyper text transport protocol
6）HTML ________________	f. hypertext markup language

3. Complete the following sentences by translating the Chinese in the brackets.

1）A browser can decode the HTML tags that are used to format pages on the Internet and can ________________（显示图像和文本）.

2）An application may ask the user if he wants to ________________（安装该程序到默认安装目录里）.

3）A good ________________（存取控制系统）should allow valid users to operate the computer easily with the minimum of checks, while barring entry to hackers or unauthorized users.

4）Windows users ________________（单击鼠标左键选定图标）and click on the right-hand button to display a menu of options that apply to the icon.

5）With a good telephone line, this pair of modems can achieve ________________（14.4Kbit/s 的传输速率）.

4. Choose the best answer for each blank.

1）The ________ is a collection of computers connected together by phone lines that allows for the global sharing information.（2005 年上半年程序员考试试题）

A. interface B. Internet C. LAN D. WWW

2）________ are those programs that help find the information you are trying to locate on the WWW.（2006 年上半年程序员考试试题）

A. Windows B. Search engines C. Web sites D. Web pages

3）A Web ________ is one of many software applications that function as the interface between a user and the Internet.（2006 年上半年程序员考试试题）

A. display B. browser C. window D. view

4）________ are Web sites that search the Web for occurrences of a specified word or phrase.（2005 年上半年程序员考试试题）

A. Search engines B. WWW C. Internet D. Java

5）________: A graphical bar with buttons that perform some of the most common commands.（2006年下半年程序员考试试题）

A. Title bar　　B. Tool bar　　C. Status bar　　D. Scroll bar

Supplementary Reading

Some Popular Browsers

Microsoft Internet Explorer

Microsoft's Internet Explorer (IE) is currently considered the dominant browser. IE offers tabbed browsing, instant access to e-mail, integrated RSS support, better standards compliance, a built-in phishing filter, advanced security (cross-domain script barriers, International Domain Name Anti-Spoofing protection and so on), and an option for deleting browsing history by removing user-identifiable information.

Mozilla Firefox Web Browser

Mozilla Firefox is a free, open source, cross-platform, Web browser developed by the Mozilla Corp. and hundreds of volunteers. Mozilla Corp. is responsible for the browser, where volunteers and enthusiasts have created many of the plug-ins available for the browser. Firefox offers support for many standards including HTML, HTML, XML, XHTML, CSS, ECMAScript (JavaScript), DTD, XSL, SVG, XPath and PNG images. The browser can also be used on a variety of operating systems such as Windows, Mac OSX, BeOS, FreeBSD, Linux, and others.

Google Chrome Web Browser

Google Chrome is a Web browser designed for Windows systems. It offers users a minimal design and what Google calls "sophisticated technology" to make the Web faster, safer, and easier on Windows-based PCs. Google Chrome features searching from the address bar, thumbnail views of your favorite pages for quick access, a private browsing function that opens an incognito window when you don't want to save your browsing history, instant bookmarks, crash control and dynamic tabs. The browser works with Windows 7 and Windows XP.

Apple Safari

Apple Safari is a Web browser available for the Macintosh and Windows operating systems as well as the iPhone, iPod Touch and iPad. Safari has been designed based on the premise that the most useful browser is one that "gets out of your way and lets you simply enjoy the Web". At the heart of Apple's Safari browser is the WebKit engine, which is responsible for functions like displaying graphics, rendering fonts, running Javascript and determining page layout.

Opera Web Browser

Opera is an international Web browser, developed in Norway. It is available for Windows XP/7 in 51 different languages. Opera includes widgets, support for BitTorrents, support for a wide variety of image, audio, and video formats, as well as enhanced HTML features, JavaScript, server push capabilities, Opera e-mail, voice technology, and client side image mapping.

New Words and Expressions

Microsoft Internet Explorer 微软 IE 浏览器
Mozilla Firefox Web browser 火狐浏览器
Google Chrome Web browser 谷歌浏览器
Apple Safari 苹果 Safari 浏览器
Opera Web browser Opera 浏览器（挪威）
BitTorrent 比特流（一种内容分配协议）
commercial [kəˈməːʃəl] *a*. 商业的，营利的，靠广告收入的
pop-up [ˈpɔpˌʌp] *a*. 弹起的，有自动起跳装置的

Reading Comprehension

Read the following statements below and decide if they are true (T) or false (F) according to the passage you have just read.

1) IE has an option for deleting browsing history by removing user-identifiable information. (　　)
2) Mozilla Firefox browser can be used on a variety of operating systems such as Windows, Mac OSX, BeOS, FreeBSD, Linux, and others. (　　)
3) At the heart of Opera Web browser is the WebKit engine, which is responsible for functions like displaying graphics, rendering fonts, running Javascript and determining page layout. (　　)
4) Opera is an international Web browser, developed by Apple corporation. (　　)
5) Google Chrome was the first commercial Web browser. (　　)

Part 3 Screen English

网络操作中的常见错误（1）

提示信息	含　义
400—Bad Request	请求失败
401—Unauthorized	未经过认证
403—Forbidden	禁止访问

（续）

提示信息	含　义
404—Not Found	没有找到该页面
500—Server Error	服务器错误
503—Service Unavailable	服务不可用
Bad File Request	错误文件请求
Socks Error	代理服务器出错
Request Timeout	连接超时
An Unexpected Web Error	不可预知的网页错误
Cannot Add Form Submission Result to Bookmark List	无法将表单结果加入书签列表
DNS① Lookup Failed	DNS 查找失败
File Contains No Data	文件无内容
Connection Refused by Host	主机拒绝连接
Viewer Not Found	找不到查看器
Unable to Locate the Server	不能定位服务器
Host Unavailable	主机不可用
Network Connection Was Refused by the Server	服务器拒绝网络连接
Host Unknown	找不到主机
Not Connect, Access Deny	拒绝连接访问
Bad Target URL	目标 URL 错误
Too Many Users	用户太多

注①：DNS 全称为“Domain Name Server”，指 Internet 上的域名服务器。

Quotation

Life is not fair; get used to it. ——Bill Gates

生活是不公平的；要去适应它。

Television is NOT real life. In real life people actually have to leave the coffee shop and go to jobs.

——Bill Gates

电视并不是真实的生活。在现实生活中，人们实际上得离开咖啡屋去干自己的工作。

Key to Exercises

Listening & Speaking

1. 1) released　2) holes　3) install　4) manually　5) apply
   6) hacker　7) addresses　8) resolved　9) font　10) component

2. 1) B 2) B 3) A 4) A 5) B 6) A

Exercises

1. 1) d 2) g 3) e 4) f 5) a 6) c 7) h 8) b
2. 1) b 万维网 2) d 统一资源标识符 3) e 超文本传输协议
   4) a 文件传输协议 5) c 因特网中继聊天（程序） 6) f 超文本标记语言
3. 1) display images and text
   2) install the application to the default
   3) access control system
   4) click on the left-hand mouse button to select an icon
   5) a transmission rate of 14.4Kbit/s
4. 1) B 2) B 3) B 4) A 5) B

Comprehension

1) T 2) T 3) F 4) F 5) F

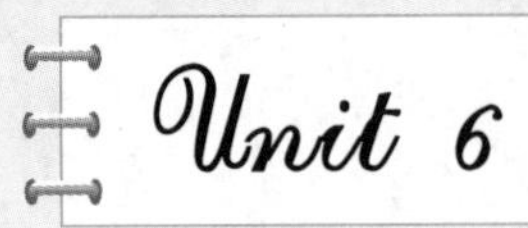

Unit 6

Multimedia

Learning Objectives

After completing this unit, you will be able to:

1. Explain how to install a webcam in English orally;
2. Identify different types of multimedia in English;
3. Know the development of sound card in English.

Part 1 Listening & Speaking

1. Listen to the following passage and fill in the blanks with the words in the box.

images	elements	selection	sound	software	modify
application	site	video	create	programs	

There is a tremendous 1) ________ of 2) ________ that can be used to 3) ________ the multimedia elements to be included in a multimedia 4) ________ or Web 5) ________. Typically this process involves several 6) ________, such as graphics software to create or 7) ________ the 8) ________, animation software to create animated 9) ________, and audio & video-editing software to create finished 10) ________ and 11) ________ clips.

2. Choose the proper words or expressions.

The joint photographic experts group 1) ________ (usually just referred to as JPEG and saved with the file extension . jpg) is supported by most Web browsers that display images. JPEG 2) ________ are compressed using lossy compression, so image quality is lost during the compression 3) ________, and a compression amount from 0 to 100 percent is selected when the image is saved—the higher the compression, the smaller the file size, but more quality is lost in the 4) ________ process. JPEG images can 5) ________ more than 16 million colors (called true color), so this format is often used for photographs and images that 6) ________ more than 256 colors.

1) A. format B. formatting
2) A. imagery B. images
3) A. process B. proceed
4) A. compressor B. compression
5) A. convert B. contain
6) A. require B. request

3. Please read the conversation below and learn how to raise a question and how to solve each problem.

网络摄像头 webcom	安装 install	摆放 lay out
磁盘 disk	插入 insert	向导 wizard
接口 connection		

A：马克，我是珍妮。

B：珍妮你好！很久没联系了，近来好吗？

A：很好。你呢？

B：不错。有什么事吗？

A：我刚买了个网络摄像头，你教我怎么安装吧。

B：好的。你先把包装里所有的东西都摆放出来，告诉我都有哪些东西。

A：只有一个摄像头和一个磁盘。

B：先插入磁盘，Windows 会识别出你要安装软件并弹出向导，指导你如向安装。

A：好了。然后呢？

B：看看那个摄像头是不是有个 USB 接口？

A：嗯。

B：把它接上（计算机），再把摄像头放在你能看到的位置。哦，应该说它能看到你的位置！

A：好了，图像出来了！

B：那就装好了。

A：太谢谢了。

B：不用谢，有问题随时来找我。再见。

A：Mark，this is Jenny.

B：Hi，Jenny. Glad to hear from you. How are you doing?

A：Pretty good. And you?

B：Not bad. Is there anything I could do for you?

A：I bought a webcam. Teach me how to fix it.

B：OK. Take out all the materials in the package，lay them out and tell me what they are.

A：Only a webcam and a disk.

B：Insert the disk，Windows will recognize that you are attempting to install software，and a wizard should pop up to guide you through the process.

A：OK. And?

B：Take a look and check if the webcam has a USB connection.

A：Yep.

B：Plug your webcam in，and you'll need to put it somewhere you can see. Oh，somewhere it can see you！

A：Wow！I can see my face！

B：OK. It's done.

A：Thank you very much.

B：Don't mention it. Call me anytime you need me. Bye.

4. Oral Practice

Mark 买了个数码照相机，想把照片复制到计算机上，请你告诉他应该如何解决这个问题。

Key Words

数码照相机 digital camera　照片 photograph/picture　传输 transmit

Part 2 Reading

Text

Multimedia

Multimedia refers to any type of application or presentation that involves integrating a variety of media, such as text, graphics, video, animation, and sound.[1] Typically, applications aren't labeled "multimedia" unless they include sound, video, animation, or interactivity.

Q1 Can books be labeled "multimedia"?

Text is an important part of most multimedia applications. It can be used to supply basic content, as well as to add text-based menus, hyperlinks, and other navigational elements. Text in application programs (including multimedia and Web development programs) can be displayed in a variety of typefaces, colors, sizes, and appearances.

Q2 How can text be used?

Graphics refers to digital representations of photographs, drawings, charts, and other visual images. Unlike animation or video, graphic images are unmoving, static images. Graphics can be created by scanning a photograph or document, taking a picture with a digital camera, or creating or modifying an image in an image-editing program.[2]

Q3 Can graphic images move?

Q4 How can graphics be created?

Animation is the term used to describe a series of graphical images that are displayed one after the other to simulate movement.[3] Cartoons on television are one example of animation. Multimedia applications, both on and off the Web, generally use some type of animation. Types of animation commonly used in

Q5 What types of animation commonly used in non-Web applications?

non-Web applications include page transitions, and animated objects, such as moving text or images. To add simple animation to a Web page, Java applets and animated GIFs are frequently used.

Q6 What types of animation used to add simple animation to a Web page?

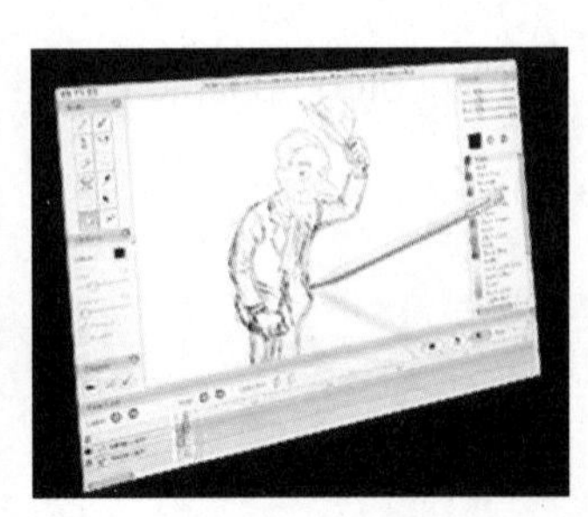

Audio includes all types of sound, such as music spoken voice, and sound effects. Most multimedia applications use sound to enhance the presentation or for content delivery. Audio for multimedia applications can be recorded using a microphone or MIDI instrument; audio can also be captured from CDs and downloaded from the Internet.

Q7 Why do multimedia applications use sound?

Q8 How can we get audio for multimedia applications?

Video differs from animation in that it usually begins as a continuous stream of visual information that is broken into separate images or frames when the video is recorded. [4] When the frames are projected—typically at a rate of 30 frames per second—the effect is a smooth reconstruction of the original continuous stream of information. Video may be recorded using a standard (analog) video camera and then converted to digital forms as it is input into a computer.

Q9 Does video differ from animation?

Q10 At what speed are the frames projected to reconstruct smooth continuous stream of information?

Q11 How is video input into a computer?

Multimedia applications appeal to a wide variety of people and can fit a variety of learning styles, and make the presented material more interesting and enjoyable, and many ideas are easier to convey in multimedia format. Although multimedia use costs much more and requires a great deal more storage space, and is limited by a variety of browsers, plug-ins and Internet connection speeds when used on the Web, the fact is that they are widely found both on and off the Web and it will be even more exciting and more embedded into everyday events than at present. [5]

Q12 What are the advantages and disadvantages of using multimedia?

New Words

integrate ['intigreit]	*v*. 使结合起来 e. g. integrate theory with practice 理论联系实际
label ['leibl]	*vt*. 标注，贴标签于 e. g. Still, the US retaliation—130 congressmen urging President Obama to label China a currency manipulator, for the first time in 16 years—is a rush of blood to the head. 不过，美方的回应——130名国会议员联名敦促奥巴马总统将中国列为汇率操纵国（这是16年来的首次）——似乎有点意气用事。
include [in'klu:d]	*vt*. 包含，包括 e. g. In any case work does not include time, but power does. 在任何情况下功都不包括时间，但功率包括时间。
supply [sə'plai]	*vt*. 供给，提供，补充 e. g. to supply communities with gas 向社区供应煤气
navigational [ˌnævi'geiʃənl]	*a*. 航行的，航运的 e. g. And they had to make this move without the navigational equipment in the command module. 而且，他们只能在指令舱没有导航设备的情况下完成这一动作。
unlike [ˌʌn'laik]	*prep*. 和……不同，不像 e. g. Smiles, unlike sandwiches, might be scarce in the snow this year. 微笑，与三明治不同，在今年的雪天里或许很稀缺。
simulate ['simjuleit]	*vt*. 模拟，模仿，仿真
spoken ['spəukən]	*a*. 口语的，口头的 e. g. ... a terminal which receives spoken information from a computer. ……一种终端，能接收来自计算机的语音信息。
effect [i'fekt]	*n*. 影响，效果，作用
enhance [in'ha:ns,-hæns]	*vt*. 提高，加强，增加 e. g. The computer can enhance the quality of photographs transmitted from space. 计算机可增强从太空发回的照片的质量。
delivery [di'livəri]	*n*. （信件、货物等的）投递，递送，传送

capture ['kæptʃə]	vt.［计］采用俘获法转换（数据），从计算机存储库中检索信息
differ ['difə]	vi.（在性质或特性上）不同，不一样，相异，有区别 e.g. Each writer's style differs from that of another. 每个作家的风格各不相同。
continuous [kən'tinjuəs]	a. 连续的，持续的
separate ['sepəreit]	a. 单独的，分开的
reconstruction [ˌriːkən'strʌkʃən]	n. 再建，重建
material [mə'tiəriəl]	n. 材料，原料
enjoyable [in'dʒɔiəbl]	a.（能）给人乐趣的，可从中得到乐趣的
convey [kən'vei]	vt. 传送，传递，传播，传导 e.g. Wires convey electricity from power stations to the users. 电线将电力从发电站传送给用户。
limit ['limit]	vt. 限制，限定 e.g. The teacher limited her to 800 words for her composition. 老师将她的作文限定在800字以内。
embed [im'bed]	vt. 把……嵌入，固定（在某物之中） e.g. So this is our little Google calendar that we embed in the course's website. 这就是被我们嵌入在课程网站中的小Google日历。

Key Terms

presentation 展示，陈述
interactivity 交互性，互动性
hyperlink 超链接
typeface 字型
photograph 照片，相片
visual 视觉的，视力的
cartoon 卡通片，［电影］动画片
Java applet 用Java语言写的小程序
animated GIF 动画GIF文件
microphone 传声器，麦克风
MIDI instrument 乐器数字接口设备
analog 模拟

Useful Expressions

not... unless 除非

e. g. You won't get paid for time off unless you have a doctor's note. 除非你有医生证明，否则你不上班就拿不到工资。

as well as 也，和……一样，不但……而且……

e. g. They sell books as well as newspapers. 他们卖书，也卖报纸。

a series of 一系列，一连串的

e. g. a random series of number 一列随机数字

both... and... 两个都，既……又……

e. g. Both his mother and his father will be there. 他的父母都会去。

all types of 形形色色的，各类

e. g. all types of software 各种软件

at a rate of 以……的速度

appeal to 引起……的好感

e. g. He doesn't appeal to her. 她对他没有好感。

Notes

1. Multimedia refers to any type of application or presentation that involves integrating a variety of media, such as text, graphics, video, animation, and sound.

that 引导定语从句，修饰 application or presentation。

2. Graphics can be created by scanning a photograph or document, taking a picture with a digital camera, or creating or modifying an image in an image-editing program.

by 引导方式状语。

3. Animation is the term used to describe a series of graphical images that are displayed one after the other to simulate movement.

that 引导定语从句，修饰 images。

4. Video differs from animation in that it usually begins as a continuous stream of visual information that is broken into separate images or frames when the video is recorded.

第一个 that 引导宾语从句和 in 一起做方式状语；第二个 that 引导定语从句，修饰 visual information。

5. ... the fact is that they are widely found both on and off the Web and it will be even more exciting and more embedded into everyday events than at present.

that 引导表语从句。

Exercises

1. Match each of the following terms with its Chinese equivalent.

1) text　　a. 储存空间
2) graphics　　b. 内容传输
3) video　　c. 图像
4) animation　　d. 动画
5) interactivity　　e. 文本
6) learning style　　f. 视频
7) storage space　　g. 交互性
8) content delivery　　h. 学习模式

2. Recognize the following abbreviations by matching them with their corresponding full names and translate them into Chinese.

1) MIDI ________________　　a. waveform audio file format
2) GIF ________________　　b. joint photographic experts group
3) JPEG ________________　　c. tagged image file format
4) TIFF ________________　　d. musical instrument digital interface
5) MPEG ________________　　e. moving picture experts group audio layer 3
6) AVI ________________　　f. graphics interchange format
7) WAV ________________　　g. motion picture experts group
8) MP3 ________________　　h. audio-video interleaved

3. Complete the following sentences by translating the Chinese in the brackets.

1) Multimedia ________________ (指任何类型的应用或展示) that involves integrating a variety of media, such as text, graphics, video, animation, and sound.

2) Text in application programs (including multimedia and Web development programs) can be displayed ________________ (用各种各样的字体、颜色、大小和外形).

3) ________________ (为了把简单的动画加入网页), Java applets and animated GIFs are frequently used.

4) Video may be recorded using a standard (analog) video camera and then converted to digital forms ________________ (当它导入计算机的时候).

5) Multimedia applications ________________ (迎合广大不同的人群) and can fit a variety of learning styles.

4. Choose the best answer for each blank. (2009 年数据库系统工程师考试试题)

For nearly ten years, the unified modeling language (UML) has been the industry standard for visualizing, specifying, constructing, and documenting the 1) ________ of a software-intensive system. As the 2) ________ standard modeling language, the UML facilitates communication and reduces confusion among project 3) ________. The recent standardization of UML 2.0 has further extended the language's scope and viability. Its inherent expressiveness allows users to 4) ________ everything from enterprise information systems and distributed Web-based applications to real-time embedded systems.

The UML is not limited to modeling software. In fact, it is expressive enough to model 5) ________systems, such as workflow in the legal system, the structure and behavior of a patient healthcare system, software engineering in aircraft combat systems, and the design of hardware.

To understand the UML, you need to form a conceptual model of the language, and this requires learning three major elements: the UML's basic building blocks, the rules that dictate how those building blocks may be put together, and some common mechanisms that apply throughout the UML.

1) A. classes B. components C. sequences D. artifacts

2) A. real B. legal C. de facto D. illegal

3) A. investors B. developers C. designers D. stakeholders

4) A. model B. code C. test D. modify

5) A. non-hardware B. non-software C. hardware D. software

Supplementary Reading

Sound Card

In the very early days, computers were silent. Then Apple company put standard sound cards and speakers in their Macintosh computers in the mid-1980s, but they were't very powerful. Still, you could hear something, and the computer could read text on a page. IBM clones really started to use more sound cards after the Creative Technologies Company of Singapore started promoting their Sound Blaster cards. Most computers now are multimedia ready.

There are many authoring software packages available for making multimedia projects now. Though some are common, there is no real standard. All these have been made possible by advances in compression technologies and transmission and storage for pictures, sound, and video. These include GIF, JPFG and TIFF for pictures, MPEG and AVI for video, and WAV and MP3 for sound.

Specialist packages for sound creation using MIDI interfaces are now available, as well as sound editing programs for working with recorded sound. Desktop video has also grown quickly, allowing cutting and pasting of video and audio clips as easily as words in a word processor. These technologies have given a big boost to the multimedia industry, as making and editing programs is now easier than ever.

New Words and Expressions

standard [ˈstændəd] *a*. 标准的

powerful [ˈpauəful] *a*. 强大的，作用大的，（机器等）大功率的

clone [kləun] *n*. 复制品

creative [kriˈeitiv] *a*. 创造性的，有创造力的

available [əˈveiləbl] *a*. 可用的，适用于……的

advance [ədˈvɑːns, ədˈvæns] *n*. 发展

compression [kəmˈpreʃən] *n*.（计算机数据的）压缩

transmission [trænzˈmiʃən, -s-] *n*. 传递，传送

specialist [ˈspeʃəlist] *n*. 专家 *a*. 专家的，专业的

processor [ˈprəusesə] *n*. [计] 处理器，处理程序，加工者

boost [buːst] *n*. 推动，帮助

Sound Blaster card 声霸卡

Reading Comprehension

Read the following statements below and decide if they are true (T) or false (F) according to the passage you have just read.

1) Computers used to be silent. ( )

2) Multimedia didn't come into being until sound cards and speakers were put in the computers. ( )

3) It was IBM that produced Sound Blaster cards. ( )

4) GIF, JPEG, TIFF, AVI, WAV and MP3 are compression technologies and transmission and storage for pictures, sound, and video. ( )

5) Cutting and pasting of video and audio clips have become easier now. ( )

Part 3 Screen English

Java 运行环境提示

提示信息	含　义
The Java Runtime Environment (JRE) The Java Runtime Environment (JRE) is available as a separately downloadable product. See the download website. The JRE allows you to run applications written in the Java programming language. Like the JDK, it contains the Java virtual machine (JVM), classes comprising the Java platform API, and supporting files. Unlike the JDK, it does not contain development tools such as compilers and debuggers. You can freely redistribute the JRE with your application, according to the terms of the JRE license. Once you have developed your application using the JDK, you can ship it with the JRE so your end-users will have a Java platform on which to run your software.	**Java 运行环境（JRE）** Java 运行环境（JRE）可作为单独的下载组件使用。参见下载网站。 JRE 允许你运行采用 Java 编程语言编写的应用程序。与 JDK 类似，JRE 含有 Java 虚拟机（JVM），构成 Java 平台 API 的类以及支持文件。JRE 与 JDK 不同，它不含如编译器、调试器的开发工具。 你可以根据 JRE 的许可条款，自由地将 JRE 与你的应用程序一起重新发布。一旦你采用 JDK 开发出了应用程序，你就能将该应用程序放在 JRE 上运行，从而你的最终用户将会拥有一个运行你的软件的 Java 平台。

Quotation

At Microsoft there are lots of brilliant ideas but the image is that they all come from the top—I'm afraid that's not quite right. ——Bill Gates

在微软，有很多高见，但看起来这些高见都来自高层——这样恐怕不太对劲。

I believe that if you show people the problems and you show them the solutions they will be moved to act. ——Bill Gates

我坚信要是给人们提出问题，并给出解决方案，人们就会行动。

Key to Exercises

Listening & Speaking

1. 1) selection　2) software　3) create　4) application　5) site

6) programs　7) modify　8) images　9) elements　10) sound
11) video

2. 1) A　2) B　3) A　4) B　5) B　6) A

Exercises

1. 1) e　2) c　3) f　4) d　5) g　6) h　7) a　8) b

2. 1) d 乐器数字接口　2) f 图像互换格式　3) b 联合图像专家小组
4) c 标签图像文件格式 5) g 动态图像专家组　6) h 音频视频交错（格式）
7) a 波形声频文件格式 8) e 动态图像专家组压缩标准音频层面 3

3. 1) refers to any type of application or presentation
2) in a variety of typefaces, colors, sizes, and appearances
3) To add simple animation to a Web page
4) as it is input into a computer
5) appeal to a wide variety of people

4. 1) D　2) C　3) D　4) A　5) B

Comprehension

1) T　2) T　3) F　4) T　5) T

Unit 7

Search Engine

Learning Objectives

After completing this unit, you will be able to:

1. Grasp the words and expressions related to this topic;
2. Master the key terms involved in this topic;
3. Be familiar with various types of search sites;
4. Gain some key searching strategies.

Part 1 Listening & Speaking

1. Listen to the following passage and fill in the blanks with the words in the box.

alterations	up-to-date	cookies	instance	indexing
regular	freshness	searcher	disadvantages	rivals

Google has overhauled the way it serves up results in response to search queries. The update is designed to work out whether a person wants 1) ________ results or historical data. The US firm estimated the 2) ________ to its core algorithm would make a difference to about 35% of searches. The changes try to make results more relevant and beef up features which Google believes set it apart from 3) ________. By contrast, Microsoft's Bing search engine emphasizes social search.

Fresh Spam

"Search results, like warm 4) ________ right out of the oven or cool refreshing fruit on a hot summer's day, are best when they're fresh," wrote Google fellow Amit Singhal in a blogpost explaining the changes.

"The under-the-hood changes sought to understand whether a 5) ________ wants results from the last week, day or even minute" said Mr. Singhal.

The update is supposed to offer a better guess of how "fresh" the results should be.

"For 6) ________," said Mr. Singhal, "anyone searching for information about the 'Occupy Oakland protests' would probably want up to the minute news."

These need to be distinguished from searches for 7) ________ events such as sports results or company reports.

"Other types of searches could call on older results," he said. "Those looking for a recipe to make tomato sauce for pasta quickly would be happy with a page that is a few months or years old."

The update to improve the "8) ________" of results builds on the big update made to the underlying infrastructure of Google's core 9) ________ system in August 2010 known as Caffeine. That change made it easier for Google to keep its index up to date and to add new sources of information.

Writing on the Search Engine Land news site, analyst Danny Sullivan described the changes as "huge". The last big update to the Google algorithm, known as Panda, affected only 12% of searches.

"The update could have potential 10) ________," warned Mr. Sullivan. "Rewarding freshness potentially introduces huge decreases in relevancy, new avenues for spamming or getting 'light' content in," said Mr. Sullivan.

2. Choose the proper words or expressions.

Dozens of popular Chinese writers have accused the search engine giant Baidu of infringing on their copyrights and branded it a "thief" in the latest claims of piracy against the company.

More than 40 writers, including the popular blogger Han Han and well-know writer Jia Pingwa, have signed a letter claiming Baidu provided their works for free download on its online library, Baidu Wenku, without their permission.

"Baidu has become a totally corrupt 1) ________ company," the authors said in the letter posted on Tuesday on the website of the China Written Works Copyright Society. "It stole our works, our rights, our property and has turned Baidu Wenku into a marketplace of stolen goods," it said.

Baidu Wenku, launched in 2009, allows users to read, share or download files and books, or their 2) ________, for free. Readers can also purchase books from the online library—at a much lower cost than the cover price.

All documents are uploaded by Internet users. As of November, Baidu Wenku had stockpiled more than 10 million files and books, accounting for 70 percent of China's online file-sharing market, according to the company's figures.

Baidu's spokesman Kaiser Kuo said the search engine "3) ________ great importance to intellectual property rights protection" and had deleted "tens of thousands of infringing items" uploaded by Web users.

"We promised that authors or copyright holders can report problematic content found on Baidu Library to the complaint center ... and we will delete infringing content within 48 hours," Kuo said in a statement on Wednesday.

In a disclaimer on its website, Baidu said users who 4) ________ the files must accept all liabilities and be responsible for compensation in any copyright disputes.

However, the writers insisted Baidu should bear responsibility, saying the company took advantage of the uploads to "enhance its own influence, 5) ________ its stock price and increase its profits."

"We do not blame the friends who uploaded (the documents). We only blame the evil platform of Baidu," they said.

Zhang Hongbo, deputy director general of the China Written Works Copyright Society, also said on Tuesday that the society is collecting more evidence and may sue Baidu for copyright violation.

1) A. thieving B. thief
2) A. except B. excerpts
3) A. attaches B. attract
4) A. uploaded B. download
5) A. boost B. boot

3. Please read the conversation below and learn how to raise a question and how to solve each problem.

查找 look for	建议 suggestion	背景资料 background information
方便 convenient	关键词 keywords	经济有效 economical and effective
主题 topic	单击 click	搜索结果 searching result
屏幕 screen	具体 specifically	考虑 consider
术语 term	同义词 synonym	

A: 嗨，汤姆，我想要查找些资料，你能给我些建议吗?

B: 当然，你要找什么资料?

A: 我们英语老师叫我们找些有关迪士尼的背景资料。

B: 我建议你上网搜索，方便又快捷。跟去图书馆一本本找书相比，你只需要输入网址或关键词就可以了。

A: 这倒是真的。网上搜索的确是既经济又有效的获取信息手段。你推荐哪些搜索网站?

B: 有不同类型的搜索网站，如百度、谷歌、雅虎，如果你要搜索英文资料，我觉得Goolge比较好。

A: 但是如何搜索呢?

B: 很简单。首先要明确你的主题，输入一些关键字，单击"搜索"。数以千计的搜索结果将出现在屏幕上。

A: 你能具体点告诉我吗? 我还是不太清楚。

B: 好的。听着，关于迪士尼这个主题，你可以输入迪士尼乐园、迪士尼人物甚至迪士尼的发展等等，也就是说要考虑所有可能的单词或词组，包括有关的术语或同义词。

A: 哦，我明白了。

B: 此外除了查看文件，图片、电影或MP3也可以在网上搜索。因特网的搜索领域已扩大到这些领域了。

A：好的。非常感谢。

B：不用客气。

A：Hi，Tom. I want to look for some information. Can you give me some advice?

B：It's my pleasure. What kinds of information are you searching for?

A：Our English teacher asked us to look for some background information about Disney.

B：I suggest searching the Internet. It's convenient and swift. Instead of going to the library and checking books one by one，the only thing you need to do is to type in the Web sites or the relevant keywords.

A：That's true. Web searching is really an effective and economical means of obtaining information. Which search sites do you recommend?

B：There are different kinds of search sites，like Baidu，Google，Yahoo!. For searching information in English，I think Goolge is better.

A：But how to search?

B：That's simple. Be clear about your topic，type some keywords and click Search. Thousands of searching result will appear on the screen.

A：Can you tell me specifically? I'm still confused.

B：OK. Listen，for this topic，besides typing Disney，you can type Disneyland，Disney characters，even the development of Disney and so on. That means you can consider all possible words or phrases，including related terms or synonyms.

A：Oh，I see.

B：In addition to looking up files，pictures，movies or MP3 can also be searched on line. Internet has expanded its searching area to these areas.

A：I agree. Thanks a million.

B：You are welcome.

4. Oral Practice

Mark 在查找有关迪士尼的资料，搜出了许多网页，但是很多都是无用的。请你告诉他如何使用 Google 高级搜索来快速查找自己想要的资料。

Key Words

访问 access	高级搜索 advanced research	精确的结果 precise result
"或" 搜索 the "OR" search	要排除的字词 unwanted words	
文件类型 file type	域 domain	

Part 2 Reading

Text

Searching the Internet

Searching the Internet is more and more popular nowadays. And how to successfully seek and locate information on the Internet is one of the most important skills an Internet user can obtain today. The Internet is a huge storehouse of interesting and useful information, but that enormous information is useless for you if you can't find it when you need it. While casual surfing is a popular Web pastime, people often turn to the Internet to find specific types of information as well. To carry out more successful Internet searches, you should be familiar with different kinds of search sites available and some key searching strategies.

Q1 Why should we be familiar with the various types of search sites and some key searching strategies?

Search Sites

There are a variety of search sites—Web sites that enable users to search for and get information on the Internet—available. Many of the most popular search sites, such as Yahoo!, Alta Vista, Go. com, Google and so on, can search using both keywords and a directory. Typically, these tools can be used interchangeably to locate Web sites including the information which you are seeking. With some exceptions, most of such search tools can be used free of charge. Since most search sites use some type of search database to locate appropriate Web pages as you search, it is important to understand a little about how such a database works.

Q2 How can the search sites search information?

Search Databases

While it may appear that a search site actually searches the Internet for you when you request it, in fact such a search would be entirely too time-consuming to perform in real time.[1] Instead, virtually all search sites use a search database previously filled with millions of URLs classified by various types of keywords or categories and a search engine program to retrieve a list of matching Web pages from the database.[2]

Q3 How does the search databases work?

Searching with Keywords

When you know generally what you want but don't know at which URL to find it, one of your best options is to perform a keyword search.[3] This type of search uses keywords (one or more key terms) that you supply to pull matching pages from its search database. Once one or more key terms are typed in the appropriate location on the search site's Web page, the search engine will retrieve and display a list of matching Web pages.

Q4 When should we need to search with keywords?

Searching with Directories

Directories are usually a good choice if you want information about a particular category, but have less of an exact subject in mind. A directory also uses a database, but that is usually screened by a human editor so it is much smaller, though often more accurate.[4] For example, while a spider may classify a page about "computer chips" under the key word "chips" together with information about potato chips. One of the largest directions—the Open Directory Project—claims to have indexed over 2 million Web pages using over 3,000 volunteer editors.

Q5 When should we need to search with directories?

To use a directory located on a search site, categories are selected instead of typing keywords. After selecting a main category, a list of more specific subcategories for the selected main category is displayed. Eventually, after selecting one or more subcategories, a list of appropriate Web pages is displayed.

Natural Language Search Sites

A new type of search site is the natural language search site. Instead of typing keywords like for a keyword search on a conventional search site, natural language search sites allow you to type the search criteria in sentence form.

Q6 What's the difference between a conventional search site and natural language search site?

Hybrid Search Sites

Most major search sites employ a few of these various search options to make their sites more versatile and useful. For example, the Yahoo! uses both keywords and a directory, and contains a variety of reference tools. The site is a natural language search site that also includes directory categories. Then there is more than one search option available, usually they

Q7 What is hybrid search sites?

can be used interchangeably, such as starting with the search engine and then selecting a directory category. Many sites automatically integrate their search options, such as displaying matching categories along with matching Web pages whenever keywords are used.

Evaluating Search Results

Once a list of potentially matching Web sites is returned as a result of a search, it is time to evaluate the sites to determine their quality and potential for meeting your needs. Two things to look for before clicking on a link for a matching page are:

Q8 What are the two things to look for before clicking on a link for a matching page?

Do the Web pages contain the exact information that you seek? Is the URL from an appropriate company or organization? For example, if you want technical specifications about a particular product, you may want to start with information on the manufacture's web site. If you are looking for government publications, stick with government Web sites.

After an appropriate Web page is found, the evaluation process is still not complete. If you are using the information on the page for something other than idle curiosity, you will want to try to be reasonably sure the information can be trusted.

New Words

storehouse [ˈstɔːhaus]	*n*. 库，仓库 e. g. The sea is the world's greatest storehouse of raw materials. 海洋是世界上最大的原料宝库。
available [əˈveiləbəl]	*a*. 现有的 e. g. Many functions available from modern software are quite flexible and have a number of options. 现代软件的许多功能相当灵活，有多个可配置的选项。
directory [diˈrektəri]	*n*. 目录 e. g. The file can be moved to a directory where it is kept for a month or so before it is physically deleted. 文件可以暂时移动到一个目录，在那里保留一个月或者保留到你物理删除它之前。
interchangeably [intəˈtʃendʒəbli]	*ad*. 可交换地 e. g. These terms can be used interchangeably. 这些术语使用时可以互换。

contain [kən'tein]	v. 包含 e. g. The View menu should contain all options that influence the way a user looks at the program's data. 视图菜单应包含影响用户查看程序数据的所有选项。
database ['deitəbeis]	n. 数据库 e. g. If an index were used as the retrieval system, the storage technique could still remain a database. 如果将索引作为检索系统使用，那么存储技术仍然只能是数据库。
appropriate [ə'prəuprit]	a. 合适的 e. g. Dialogs are appropriate for functions that are out of the main interaction flow. 对话框适合那些主交互流之外的功能。
virtually ['vəːtjuəli]	ad. 实际上，事实上 e. g. Clearly, this is virtually never the case in a microprocessor system. 显然，这种情况在微处理器系统中实际上是永远不会出现的。
previously ['priːviəsli]	ad. 以前，过去 e. g. The output for the problem previously conducted numerically will now be given. 下面给出前面用数值所处理的问题的输出。
retrieve [ri'triːv]	v. 检索 e. g. Whenever a record was entered, the program would return a token that could be used to retrieve the record. 无论什么时候输入记录，程序都将返回一个用于检索记录的标志。
option ['ɔpʃən]	n. 选项 e. g. The program didn't offer that option, so the signs were posted. 但是程序却没有提供该选项，所以显示了文字提示。
versatile ['vəːsətail]	a. 多才多艺的，有多种用途的 e. g. This is an extremely versatile new kitchen machine. 这是一种功能极多的新型厨房用机器。
automatically [ˌɔːtə'mætikli]	ad. 自动地 e. g. Applications should automatically save documents. 程序应该自动保存文档。
integrate ['intigreit]	vt. 使成一体，使结合，使合并 e. g. To integrate directories, IS managers must select a directory standard from among many choices. 为了把目录集成在一起，信息系统的管理人员必须从多种选择中选一个目录标准。

potentially [pə'tenʃəli]	*ad*. 潜在地，可能地 e. g. By modifying the URL, attackers can reverse-engineer the database structure and potentially find users' names, passwords, or even credit card numbers. 通过修改 URL，攻击者可以对数据库结构逆向开发，有可能找到用户姓名、密码甚至信用卡号。
evaluate [i'væljueit]	*v*. 评估，评价 e. g. How would you evaluate it as an outsider? 作为一个局外人，你对此如何评价?
specification ['spesifi'keiʃən]	*n*. 规格，明细单 e. g. All these parts were machined strictly to specification. 所有这些零部件都是严格地按技术要求加工制造的。
publication [ˌpʌbli'keiʃən]	*n*. 发表，宣布，公布 e. g. How long is it between the date of publication and the date of approval? 从公告那天起到核准那天止要多长时间?
idle ['aid(ə)l]	*a*. 无根据的 e. g. The above assertion is not idle speculation. 以上的推断并非毫无根据。
versatile ['vəːsətail]	*a*. 通用的 e. g. In this article, you learned how to create a very versatile OBEX client application. 在这篇文章中，我们学习了如何创建通用的 OBEX 客户机应用程序。
curiosity [ˌkjuəri'ɔsiti]	*n*. 好奇心 e. g. His curiosity prompted him to ask questions. 他的好奇心驱使他提问。

Useful Expressions

be familiar with 熟悉

e. g. Our technical development may be familiar with the problem. 我们的技术部门也许熟悉这个问题。

vary with 随……而变化

e. g. Details of how this is done vary with the platform and personal configuration. 由于平台和个人配置的不同，如何这样做的细节也不同。

supply to 提供给

e. g. I hope you'll increase the supply to meet the grow demand at our end. 希望贵方增加供

货以满足我方日益增长的需求。

instead of 而不是，代替

e. g. Instead of software rejecting input，it must work harder to understand and reconcile confusing input. 软件不但不能拒绝输入，而且必须努力理解和调整输入造成的混乱。

a list of 一列，一栏，一份……的单子

e. g. Each underlined phrase，when clicked，provides a drop-down menu with a list of selectable options. 当单击每个带下划线的短语时，程序都会提供一个下拉菜单，它带有一个可选择的选项列表。

stick with 跟随，坚持做

e. g. But for most products，we recommend that you stick with more basic mouse actions. 对于绝大多数产品，我们建议你还是坚持使用更基本的一些鼠标动作。

Notes

1．While it may appear that a search site actually searches the Internet for you when you request it，in fact such a search would be entirely too time-consuming to perform in real time.

该句中 while 意为"虽然，尽管"，引导让步状语从句，that 引导宾语从句，在此无意义，when 引导条件状语从句。

2．Instead，virtually all search sites use a search database previously filled with millions of URLs classified by various types of keywords or categories and a search engine program to retrieve a list of matching Web pages from the database.

该句子中 filled with millions of URLs 做后置定语修饰 a search database，而 classified by various types of keywords or categories 则修饰 URLs。

3．When you know generally what you want but don't know at which URL to find it，one of your best options is to perform a keyword search.

在此句中 when 作条件状语，what you want 充当 know 的宾语从句。

4．A directory also uses a database，but that is usually screened by a human editor so it is much smaller，though often more accurate.

此句中 that 指代前面那句话。

Exercises

1. Match each of the following terms with its Chinese equivalent.

1) database	a. 评估搜索结果
2) subcategory	b. 子类
3) specification	c. 数据库

4) retrieve　　d. 检索
5) search sites　　e. 搜索网站
6) searching with keywords　　f. 规格
7) hybrid search sites　　g. 混合查找
8) searching with directories　　h. 自然语言查找
9) evaluating search results　　i. 目录查找
10) natural language search　　j. 关键字搜索

2. Recognize the following abbreviations by matching them with their corresponding full names and translate them into Chinese.

1) WWW ________________　　a. uniform resource location
2) HTTP ________________　　b. Internet service provider
3) URL ________________　　c. World Wide Web
4) ISP ________________　　d. hyper text transfer protocol
5) DFS ________________　　e. breadth first search
6) BFS ________________　　f. depth first search

3. Complete the following sentences by translating the Chinese in the brackets.

1) ________________ (通常，这些工具允许交互式使用) to find Web sites containing the information that you are seeking.

2) When you know generally what you want but don't know at which URL to find it, ________________ (最好使用关键字查找).

3) Directories are usually a good choice ________________ (如果想知道一个特定类别的信息) but have less of an exact subject in mind.

4) If you are looking for government publications, ________________ (则关注政府网站).

5) Most major search sites employ a few of these various search options to be ____________ (更加通用和有效).

4. Choose the best answer for each blank. (2008年下半年程序员考试上午试题 (B))

1) ________ is a device that converts images to digital format.

A. Copier　　B. Printer　　C. Scanner　　D. Display

2) In C language, a ________ is a series of characters enclosed in double quotes.

A. matrix　　B. string　　C. program　　D. stream

3) ________ are those programs that help find the information you are trying to locate on the WWW.

A. Windows　　B. Search engines　　C. Web sites　　D. Web pages

4) In C language, ________ are used to create variables and are grouped at the top of a

gram block.

A. declarations B. dimensions C. comments D. descriptions

5) An ________ statement can perform a calculation and store the result in a variable so that it can be used later.

A. executable B. input C. output D. assignment

6) Each program module is compiled separately and the resulting ________ files are linked together to make an executable application.

A. assembler B. source C. library D. object

7) ________ is the conscious effort to make all jobs similar, routine, and interchangeable.

A. WWW B. Informatization C. Computerization D. Standardization

8) A Web ________ is one of many software applications that function as the interface between a user and the Internet.

A. display B. browser C. window D. view

9) Firewall is a ________ mechanism used by organizations to protect their LANs from Internet.

A. reliable B. stable C. peaceful D. security

10) A query is used to search through the database to locate a particular record or records, which conform to specified ________.

A. criteria B. standards C. methods D. conditions

Supplementary Reading

The Best Search Engines of 2011

Most people don't want 290 search engines, especially people who are Internet beginners. Most users want a single search engine that delivers three key features:

1. Relevant results (results you are actually interested in);
2. Uncluttered, easy to read interface;
3. Helpful options to broaden or tighten a search.

With this criteria, 5 Reader Favorite Search Engines come to mind.

DuckDuckGo

At first, DuckDuckGo. com looks like Google. But there are many subtleties that make this spartan search engine different. DuckDuckGo has some slick features, like "zero-click" information (all your answers are found on the first results page). DuckDuckGo offers disambiguation prompts (helps to clarify what question you are really asking). And the ad spam is much less than Google. Give DuckDuckGo. com a try and you might really like this clean and simple

search engine.

Ask (aka "Ask Jeeves")

The Ask/AJ/Ask Jeeves search engine is a longtime name in the World Wide Web. The super-clean interface rivals the other major search engines, and the search options are as good as Google or Bing or DuckDuckGo. The results groupings are what really make Ask. com stand out. The presentation is arguably cleaner and easier to read than Google or Yahoo! or Bing, and the results groups seem to be more relevant. Decide for yourself if you agree and give Ask. com a whirl, and compare it to the other search engines you like.

The Internet Archive

The Internet Archive is a favorite destination for longtime Web lovers. The Archive has been taking snapshots of the entire World Wide Web for years now, allowing you and me to travel back in time to see what a web page looked like in 1999, or what the news was like around Hurricane Katrina in 2005. You won't visit the Archive daily, like you would Google or Yahoo! or Bing, but when you do have need to travel back in time, use this search site.

Yahoo!

Yahoo! is several things: it is a search engine, a news aggregator, a shopping center, an emailbox, a travel directory, a horoscope and games center, and more. This "Web portal" breadth of choice makes this a very helpful site for Internet beginners. Searching the Web should also be about discovery and exploration, and Yahoo! delivers that in wholesale quantities.

Webopedia

Webopedia is one of the most useful websites on the World Wide Web. Webopedia is an encyclopedic resource dedicated to searching techno terminology and computer definitions. Teach yourself what "domain name system" is, or teach yourself what "DDRAM" means on your computer. Webopedia is absolutely a perfect resource for non-technical people to make more sense of the computers around them.

New Words and Expressions

uncluttered [ˌʌn'klʌtəd] *a*. 不凌乱的，整齐的

interface ['intəfeis] *n*. 界面，分界面

subtlety ['sʌtlti] *n*. 细微的差别，微妙之处

spartan ['spɑːtn] *a*. 简朴的

slick [slik] *a*. 华而不实的

disambiguation [ˌdisæmbigju'eiʃən] *n*. 消除歧义

spam [spæm] n. 垃圾电子邮件
arguably ['ɑːgjuəbli] ad. 雄辩地，可以认为
whirl [hwəːl] n. 尝试
snapshot ['snæpʃɔt] n. 简要印象，点滴的了解
aggregator ['ægrigeitə(r)] n. 聚合器，一种信息处理工具
horoscope ['hɔrəskəup] n. 占星术，星象
portal ['pɔːtl] n. 门户网站
breadth [bredθ] n. 广泛性，广度
encyclopedic [enˌsaiklə'piːdik] a. 百科全书的
terminology [ˌtəːmi'nɔlədʒi] n. (总称) 术语，专门用语

Reading Comprehension

Read the following statements below and decide if they are true (T) or false (F) according to the passage you have just read.

1) DuckDuckGo differs from Google only because it has the feature of "zero click" information. ( )
2) The presentation of Ask/AJ/Ask Jeeves search engine is cleaner and easier. ( )
3) Yahoo! is only an email box. ( )
4) The Internet Archive's Web page looks like in 1999. ( )
5) If you want to search techno terminology and computer definitions, Webopedia is a good choice. ( )

Part 3 Screen English

标准 CMOS 参数设置

项　目	参　数
Date (mm: dd: yy)　日期 (月份: 日期: 年份)	
Time (hh: mm: ss)　时间 (时: 分: 秒)	
Primary Master　第一主硬盘 Primary Slave　第一从硬盘 Secondary Master　第二主硬盘 Secondary Slave　第二从硬盘	1～45, Auto, User, None　硬盘类型 (1～45), 自动, 用户定义, 无硬盘
Drive A/B　A/B 驱动器	360K, 5.25 in/1.2M, 5.25 in/720K, 3.5 in/1.44M, 3.5 in/2.88M, 3.5 in/None
Video　视频	EGA/VGA, CGA, MONO

（续）

项　目	参　数
Halt On① 在什么情况下中断	All Errors 所有错误都中断 No Errors 所有错误都忽略 All, But keyboard 除了键盘错误，其余错误中断 All, But Diskette 除了磁盘错误，其余错误中断 All, But Disk/key 除了键盘和磁盘错误，其余错误中断

注①："Halt On"表示系统开机启动过程中，在何种情况下中断启动过程，相应的参数列出了各种可能的情况。

Quotation

The world won't care about your self-esteem. The world will expect you to accomplish something before you feel good about yourself. —— Bill Gates

这世界并不会在意你的自尊。这世界希望你在自我感觉良好之前先要有所成就。

If you mess up, it's not your parents' fault, so don't whine about our mistakes, learn from them.

—— Bill Gates

如果你陷入困境，那不是你父母的过错，所以不要尖声抱怨我们的错误，要从中吸取教训。

Key to Exercises

Listening & Speaking

1. 1) up-to-date　2) alterations　3) rivals　4) cookies　5) searcher
   6) instance　7) regular　8) freshness　9) indexing　10) disadvantages
2. 1) A　2) B　3) A　4) A　5) A

Exercises

1. 1) c　2) b　3) f　4) d　5) e　6) j　7) g　8) i　9) a　10) h
2. 1) c 万维网　2) d 超文本传输协议　3) a 统一资源定位
   4) b 因特网服务提供商　5) f 深度优先搜索法　6) e 广度优先搜索法
3. 1) Typically, these tools can be used interchangeably
   2) one of the best options is to perform a keyword search

3) if you want information about a particular category

4) stick with the government Web sites

5) more versatile and useful

4. 1) C 2) B 3) B 4) A 5) D 6) D 7) D 8) B 9) D 10) A

Comprehension

1) F 2) T 3) F 4) F 5) T

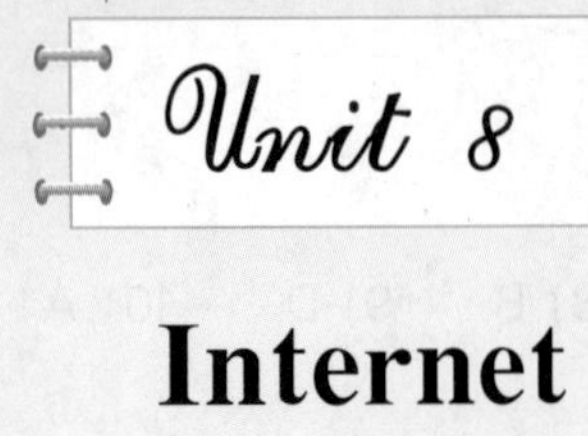

Internet

Learning Objectives

After completing this unit, you will be able to:

1. Explain how to sign up a free QQ ID in English orally;
2. Identify the components of Internet community in English;
3. Know the different types of topology of a network in English.

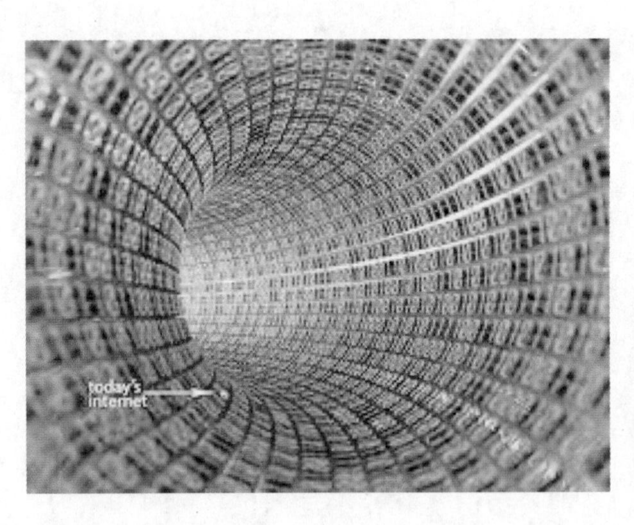

Part 1 Listening & Speaking

1. Listen to the following passage and fill in the blanks with the words in the box.

networks	Internet	connected	world	part	hyperlinks
pages	through	computers	sound	text	collectiion

The 1) ________ is a collection of 2) ________ that connects millions of 3) ________ all over the 4) ________. The World Wide Web or Web is one 5) ________ of the Internet. It is a 6) ________ of documents—called Web 7) ________ —that are accessed 8) ________ the Internet. Web pages are 9) ________ to each other by 10) ________ and commonly contain a variety of 11) ________, images, and animated objects; they can contain 12) ________ and video clips, as well.

2. Choose the proper words or expressions.

Networks often must communicate with 1) ________ resources, such as those on other networks. Messages 2) ________ between two distinct networks reach their destinations via gateways and bridges.

A gateway is a collection of hardware and software 3) ________ that enable devices on one network to communicate with those on another, dissimilar network. Computers on a LAN, for instance, require a gateway to 4) ________ the Internet. Two networks based on similar technology—such as a LAN in one city and similar LAN in another—communicate via a device called a bridge. Bridges can be used to partition one large LAN into two smaller ones.

Gateway

1) A. outside B. inside

2) A. send B. sent

3) A. resolutions B. resources

4) A. access B. accept

3. Please read the conversation below and learn how to raise a question and how to solve each problem.

申请 sign up	免费 free	向导 wizard
加某人为好友 add sb. as a friend		

A：马克，是我，珍妮。

B：珍妮！你好吗？

A：很好。你呢？

B：还不错。又遇到什么问题了？

A：我想申请个免费的QQ号。

B：好的。你先打开一个搜索引擎，然后输入“申请免费QQ号”。你会看到很多相关网页，找到那个可以注册的网页，然后按照向导完成就好。

A：知道了！我先试试。

B：哦，申请到QQ号后，它会提示你下载QQ，安装后，你才能用哦。

A：明白。

B：不明白再给我打电话。记得加我为好友哦！

A：当然！再见！

B：再见！

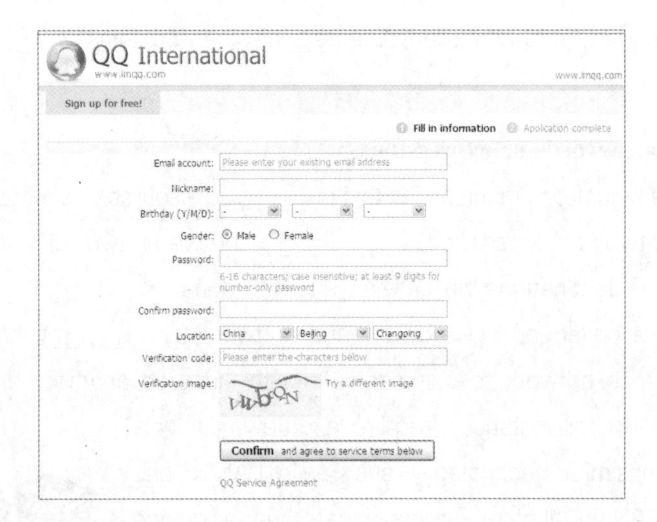

A: Mark, it's Jenny.

B: Jenny! How are you?

A: Fine. Thanks, and you?

B: So-so. So, what's the problem?

A: I want to sign up a free QQ ID.

B: OK. Open a search engine, and enter "sign up for a free QQ ID", and there are a long list of relative websites. Click some, you'll find the right one, do as the wizard tells.

A: OK. I'll try myself.

B: Oh, after that, you'll be reminded of downloading QQ. After you install it, you can sign in.

A: Got it.

B: Call me any time. Do remember add me as a friend.

A: Sure. Bye!

B: Bye!

4. Oral Practice

Jenny 想申请一个支付宝账号，请帮她解决这个问题。

Key Words

注册支付宝账号 sign up for an Alipay account　　密码 password

Part 2 Reading

Text

The Internet Community Today

The Internet community today is populated by individuals, companies, and a variety of organizations located throughout the world.[1] Virtually anyone with a computer that has communication capabilities can be part of the Internet, either as a user or as a supplier of information or services.[2] Most members of the Internet community fall into one or more of the following groups.

Q1 What can be part of the Internet?

Users are people who avail themselves of the vast amount of resources available through the Internet at work or in their personal lives. There were up to 404 million Internet users in China by the end of April in 2010. The Internet use begin to approach the popularity of the phone and TV.

Q2 How many Internet users are there in China by the end of April 2010?

Internet service providers (ISPs)—often called service providers or access providers for short—are organizations that provide Internet access to others, such as China Telecom, China Unicom, China Mobile. They operate very much like a cross between cable-television and phone companies in that they provide access to a communications service, usually for a monthly fee.[3]

Q3 Could you list some ISPs in China?

Internet content providers, or content providers, are the parties that furnish the information available through the Internet.

Q4 As an Internet content provider, what does a software company do? What about a music publisher and a film student?

Here are some examples of content providers:

- A software company creates a Web site that users can access to both get product information and download trial copies of software.[4]
- A music publisher creates a site where sample songs can be downloaded and custom CDs can be created, purchased, and downloaded.[5]
- A film studio releases its original short movies to be viewed on the Web.

Application service providers (ASPs) are companies that manage and distribute software-based services and solutions to customers across a network—usually the Internet.[6] Instead of providing access to the Internet like ISPs do, ASPs provide access to software applications. In essence, ASPs rent software access to companies or individuals. Just as with Internet access, customers typically pay a monthly or yearly fee to use the applications.

Q5 *How does ASPs differ from ISPs?*

The Internet links users throughout the world with the help of ISPs, Internet content providers, ASPs, and creates cyberspace—a non-physical "space", an Internet community without boundary.

New Words

populate ['pɔpjuleit]	*vt*. 占据，(事物等) 在……中占有位置 e. g. Her paintings were populated by strange creatures. 她的画中多是奇怪的生物。
individual [ˌindi'vidʒuəl]	*n*. 个人，个体
organization [ˌɔːgənai'zeiʃən]	*n*. 组织，机构
locate [ləu'keit]	*vt*. 位于，使……坐落在 (常用被动语态) e. g. The school is located next to the church. 学校紧挨着教堂。
throughout [θruː'aut]	*prep*. 贯穿，遍及 e. g. He searched throughout the room. 他搜遍了整个屋子。

virtually ['vəːtʃuəli]	*ad*. 实际上，事实上，实质上
avail [ə'veil]	*vt*. 有益于
approach [ə'prəutʃ]	*vt*. (在性质、数量、质量等方面) 近似，近于，接近 e. g. As a poet he hardly approaches John Milton. 同为诗人，他很难与约翰·米尔顿相媲美。
monthly ['mʌnθli]	*a*. 每月一次的
furnish ['fəːniʃ]	*vt*. 供应，提供 e. g. to furnish information 提供情报
distribute [di'stribjuːt, 'dis-]	*vt*. 分配，分发 e. g. The relief agency will distribute the food among several countries. 救援机构将给几个国家发放食物。

Key Terms

the Internet community 网上社区，网络社区
user 用户
supplier 供应商
access [计算机] 访问，存取，进网
cable-television 有线电视
communications service 通信服务
trial copy 体验版
release 发布，发行
ISP 因特网服务提供商
Internet content providers 因特网信息提供商
ASP 应用服务提供商
syberspace (cyberspace) ['saibəˌspeis] (科幻小说中) 由计算机网络组成的空间，虚拟现实 (= virtual reality)

Useful Expressions

either... or... 二者择一的，要么……要么……

e. g. I'm going to buy either a camera or a CD player with the money.

fall into 落入，分成

e. g. These problems may fall into three classes. 这些问题可以分成三类。

avail oneself of... 利用，趁 (机会)，借此 (机会)

e. g. I should avail myself of every opportunity to practice speaking English. 我应该利用一切机会练习说英语。

instead of 代替，不是……而是……

e. g. They can now spend their energy on creating new ideas instead of reinventing ideas that other people have already done quite well. 他们现在可以创造自己的观点，而不用重新整理别人已经阐述得很好的观点。

in essence 本质上，其实，大体上

e. g. In essence, your situation isn't so different from mine. 从本质上讲，你我的情况并不是相差很远。

Notes

1. The Internet community today is populated by individuals, companies, and a variety of organizations located throughout the world.

located throughout the world 过去分词短语做 individuals, companies, and a variety of organizations 的后置定语，表被动。

2. Virtually anyone with a computer that has communication capabilities can be part of the Internet, either as a user or as a supplier of information or services.

介词短语 with a computer 修饰 anyone, that 引导的定语从句修饰 computer。

3. Internet service providers (ISPs) —often called service providers or access providers for short—are organizations that provide Internet access to others, such as China Telecom, China Unicom, China Mobile. They operate very much like a cross between cable-television and phone companies in that they provide access to a communications service, usually for a monthly fee.

第一句中 that 引导定语从句修饰 organizations，第二句中 that 引导宾语从句和 in 一起做状语。

4. A software company creates a Web site that users can access to both get product information and download trial copies of software.

此句中 that 引导非限制性定语从句修饰 a Web site.

5. A music publisher creates a site where sample songs can be downloaded and custom CDs can be created, purchased, and downloaded.

where 引导定语从句修饰 a site。

6. Application service providers (ASPs) are companies that manage and distribute software-based services and solutions to customers across a network—usually the Internet.

that 引导定语从句修饰 companies。

Exercises

1. Match each of the following terms with its Chinese equivalent.

1) the Internet community	a. 路径
2) supplier	b. 网络空间
3) access	c. 网上社区，网络社区
4) cable-television	d. 因特网信息提供商
5) trial copy	e. 应用服务提供商
6) ISP	f. 因特网服务提供商
7) Internet content providers	g. 有线电视
8) ASP	h. 体验版
9) syberspace	i. 供应商

2. Recognize the following abbreviations by matching them with their corresponding full names and translate them into Chinese.

1) ISP ________________	a. personal area network
2) ASP ________________	b. Internet service provider
3) PAN ________________	c. metropolitan area network
4) NAN ________________	d. wide area network
5) LAN ________________	e. local area network
6) WAN ________________	f. application service provider
7) MAN ________________	g. neighborhood area network

3. Complete the following sentences by translating the Chinese in the brackets.

1) Virtually anyone with a computer that has communication capabilities can be part of the Internet, ________________ (无论是作为用户还是信息供应商还是服务供应商).

2) ________________ (网上社区的大部分成员) fall into one or more of the following groups.

3) Internet service providers (ISPs) are ________________ (给他人提供因特网接入服务的机构).

4) ________________ (在本质上), ASPs rent software access to companies or individuals.

4. Choose the best answer for each blank. (2010 年 5 月网络管理员考试上午试题)

Both bus and tree topologies are characterized by the use of multipoint 1) ________. For the bus, all stations attach, through appropriate hardware 2) ________ known as a tap, directly to a linear transmission medium, or bus. Full-duplex operation between the station and the tap allows data to be transmitted onto the bus and received from the 3) ________. A transmission from any station propagates the length of the medium in both directions and can be received by all other 4) ________. At each end of the bus is a 5) ________, which absorbs any signal, removing it from the bus.

1) A. medium B. connection C. token D. resource
2) A. processing B. switching C. routing D. interfacing
3) A. tree B. bus C. star D. ring
4) A. routers B. stations C. servers D. switches
5) A. tap B. repeat C. terminator D. concentrator

Supplementary Reading

Networks

A network consists of a collection of computers and other hardware devices that are together to share hardware, software, and data, as well as to facilitate electronic communications.

From a geographic perspective, networks can be classified as PANs, NANs, LANs, MANs, and WANs.

Three common topologies are the star, bus, and ring. The pathways shown between nodes can be linked by physical cables or wireless signals.

The arrangement of devices in a network is referred to as its physical topology. A network arranged as a **star topology** features a central connection point for all workstations and peripherals. The central connection point is not necessarily a server—more typically it is a network device called a hub, which is designed to broadcast data to workstations and peripherals.

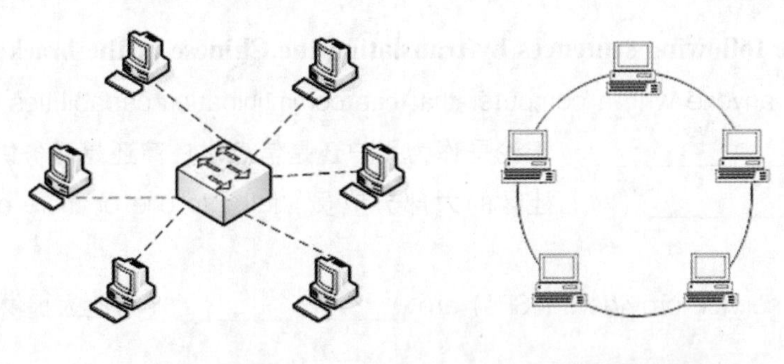

Star Topology **Ring Topology**

A **ring topology** connects all devices in a circle, with each device having exactly two neighbors. Data is transmitted from one device to another around the ring. This topology minimizes cabling, but failure of any one device can take down the entire network. Ring topologies, once championed by IBM, are infrequently used in today's networks.

A **bus topology** uses a common backbone to connect all network devices. The backbone functions as a shared communication link, which carries network data. The backbone stops at each end of the network with a special device called a "terminator". However, bus networks work best with a limited number of devices. Bus networks with more than a few dozen computers are likely to perform poorly, and if the backbone cable fails, the entire network becomes unusable.

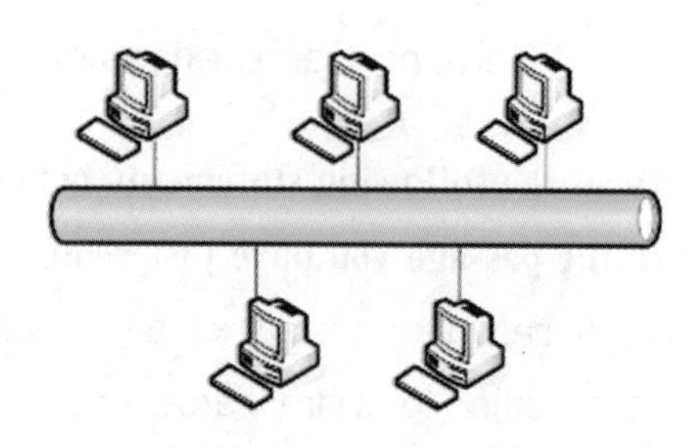

Bus Topology

A **mesh topology** connects each network device to many other network devices. Data traveling on a mesh network can take any of several possible paths from its source to its destination. Even if several links fail, data can follow alternative functioning links to reach its destination—an advantage over networks arranged in a star topology. The original plan for the Internet was based on mesh topology.

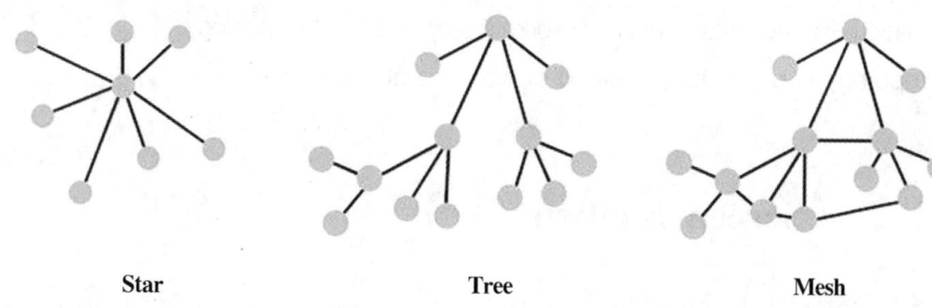

Star **Tree** **Mesh**

A **tree topology** is essentially a blend of star and bus networks. Multiple star networks are connected into a bus configuration by a backbone. Tree topologies offer excellent flexibility for expansion—a single link to the backbone can add an entire group of star-configured devices. This link can be accomplished using the same type of hub used as the central connection point in a star network. Most of today's school and business networks are based on tree topologies.

New Words and Expressions

facilitate [fə'siliteit] *vt*. 促进，帮助，使容易
geographic [ˌdʒiə'græfik] *a*. 地理的，地理学的
perspective [pə'spektiv] *n*. 角度
topology [tə'pɔlədʒi] *n*. 拓扑
feature ['fiːtʃə(r)] *vt*. 以……特征
necessarily ['nesəsərəli, ˌnesə'serəli] *ad*. 必要地，必需地
broadcast ['brɔːdkɑːst] *vt*. 播送，广播
minimize ['minimaiz] *vt*. 使减缩到最低
share [ʃεə(r)] *vt*. 分享，分担，分配
poorly ['puəli] *ad*. 不足，很差，不充分地
fail [feil] *vi*. 失灵，失败
unusable [ˌʌn'juːzəbl] *a*. 不可用的
mesh [meʃ] *n*. 网状结构
alternative [ɔːl'təːnətiv] *a*. 可供选择的
original [ə'ridʒənl] *a*. 原来的
blend [blend] *n*. 混合物（体）
configuration [kənˌfigju'reiʃən] *n*. 布局，构造
offer ['ɔfə(r)] *vt*. 提供
flexibility [ˌfleksi'biliti] *n*. 灵活性
accomplish [ə'kʌmpliʃ] *vt*. 完成（任务）

Reading Comprehension

Read the following statements below and decide if they are true (T) or false (F) according to the passage you have just read.

1) A network consists of a collection of computers and other software devices that are together to share hardware, software, and data, as well as to facilitate electronic communications. (　　)

2) A network arranged as a ring topology features a central connection point for all workstations and peripherals. (　　)

3) A bus topology uses a common backbone to connect all network devices. (　　)

4) In a mesh topology, even if several links fail, data can follow alternative functioning links to reach its destination—an advantage over networks arranged in a star topology. (　　)

5) A tree topology is essentially a blend of star and bus networks. (　　)

Part 3 Screen English

Windows XP SP2 的安装

提示信息	含　义
Windows XP Service Pack 2, offering many security and performance upgrades for your operating system, and can be conveniently downloaded in several ways. A) Visit Windows Update using Internet Explorer. Just choose Tools then Windows Update. Follow the on-screen prompts to download the product. B) Turn on the Automatic Update feature. Do this, and Windows XP will automatically download Service Pack 2 for you in the background while you are online. 1) Click Start, choosing Control Panel. 2) Click Performance and Maintenance (if that is not visible, ignore this step). 3) Choose System. 4) When the System multi-tabbed dialog box appears, click the Automatic Updates tab. 5) Choose to be notified about updates,	Windows XP SP2 能为现有操作系统提供许多安全与性能上的功能升级，可用多种方法方便地下载。 A) 使用 IE 浏览器访问 Windows 更新站点。仅需选择“工具”，然后是“Windows 更新”。按照屏幕提示下载该产品。 B) 打开“自动更新”功能。这样当你处于联机状态时，Windows XP 会自动以后台方式下载 SP2。 1) 单击“开始”按钮，选择“控制面板”。 2) 单击“性能和维护”(若该选项不可见，则可忽略该步骤)。 3) 选择“系统”。 4) 当出现“系统”的多选项卡对话框时，单击“自动更新”选项卡。 5) 选择是否需要更新通告、是否以后

（续）

提示信息	含　义
download updates in the background and be notified when they are ready to download, or automatically download the updates and install them at a specified time. 6) Click OK to close the dialog box.	台方式下载更新，以及准备好下载时是否通告，或者在指定时间自动下载更新并进行安装的选项。 6）单击“确定”按钮关闭对话框。

Quotation

I think the Internet is uniquely suited to this free market idea: that everyone on the Internet that exchanges the traffic back and forth, big or small, we all need each other.

——Pete Ashdown

我认为因特网特别适用自由贸易的理念：每位用户在互联网上来回交换流量，无论大小，只要我们彼此需要。

On the Internet, nobody knows you're a dog.

——Peter Steiner

在因特网上，没人知道你是条狗。

Key to Exercises

Listening & Speaking

1. 1) Internet　2) networks　3) computers　4) world　5) part
   6) collection　7) pages　8) through　9) connected　10) hyperlinks
   11) text　12) sound
2. 1) A　2) B　3) B　4) A

Exercises

1. 1) c　2) i　3) a　4) g　5) h　6) f　7) d　8) e　9) b
2. 1) b 因特网服务提供商　2) f 应用服务提供商
   3) a 个人局域网　4) g 邻域局域网　5) e 本地局域网
   6) d 广域网　7) c 城域网
3. 1) either as a user or a supplier of information or services
   2) Most members of the Internet community
   3) organizations that provide Internet access to others
   4) In essence
4. 1) A　2) D　3) B　4) B　5) C

Comprehension

1) F　2) F　3) T　4) T　5) T

Unit 9

Tools for Online Communication

Learning Objectives

After completing this unit, you will be able to:

1. Identify different tools for online communication;
2. Know how to use these tools.

Blog

VoIP

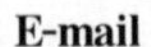

E-mail

Video Conference

Part 1 Listening & Speaking

1. Listen to the following passage and fill in the blanks with the words in the box.

talk	type	internationally	Internet	tool	act
visual	telephone	available	share	visual	

The question comes to mind, why would you use VoIP in place of the 1) ________? And the first thought is, VoIP works with a computer and an 2) ________ connection and it allows you to make phone calls to anyone else who has a computer for zero cost.

You can talk to anyone locally, anyone interstate or anyone 3) ________ by using VoIP. Every computer has the potential to 4) ________ as a telephone. If you were in a school and all the phone lines were being used, then your wireless connection computer potentially could be used to make those calls. VoIP also allows you to 5) ________ material on the screen and to 6) ________ at the same time but also be a hands free environment so it makes it a lot easier to 7) ________ when you're not holding a handset. Programs such as Bridgit are freely 8) ________ for DET. They use VoIP when you use desktops or laptops. It is also used for Web conferencing software. Skype is another popular 9) ________ and it is one of many that use VoIP for talking with the addition of a webcam that could also allow you to have 10) ________ images of the person on the other end.

2. Choose the proper words or expressions.

How to Use Blogger to Add an Image to Your Blog

One way to make your blog post more appealing is to include an image. Blogger has built-in tools for uploading a photograph or piece of artwork that's already the right size and format for displaying on the Web.

1. Click the Add Image icon.

It looks like a photograph. The 1) ________ Images window opens. Some Web browsers have begun to include functionality that blocks pop-up browser windows from opening. If you click the Add Image 2) ________ and nothing happens, go into your browser settings to turn off this protection. In some browsers, it's possible to do this for certain Web sites rather than for all of them, so you can choose to allow Blogger to open a pop-up window without subjecting yourself to other annoying pop-ups. Consult your browser's Help menu if you need assistance doing this.

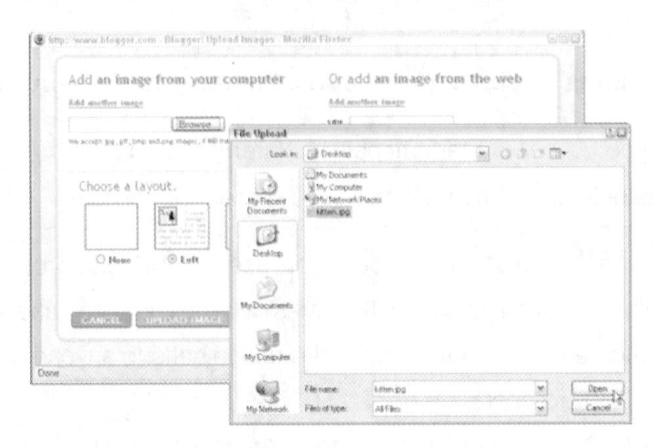

2. Click the Browse button in the Add an Image from Your Computer section of the page.

A File Upload dialog box opens.

3. Locate the image you want to upload from your computer, select it, and click Open.

The location of the image is 3) ________ into the image field.

4. Choose a layout and image size from the options.

Layout 4) ________ how text will wrap around the image. You can choose None, Left, Center, or Right. Image size determines what size the image will be shown in your blog post, regardless of how big the source image is. You can choose Small, Medium, or Large.

5. Click to accept the Terms of Service and then click the Upload Image button.

Your image is uploaded and inserted into your blog post.

6. Click Done.

The Upload Images window closes, and your image is 5) ________ in your blog post field.

1) A. Upload　　B. Download
2) A. icon　　B. figure
3) A. injected　　B. inserted
4) A. determines　　B. detect
5) A. noticeable　　B. visible

3. Please read the conversation below and learn how to raise a question and how to solve each problem.

因特网 Internet	介绍 introduce	免费的 free
推荐 recommend	好用的 easy-to-use	联系 contact
个人的 personal	建立 create	账号 account
发布 publish	模板 template	

A: 有什么可以为你效劳的吗?

B: 我想在因特网上和朋友进行交流，能不能给我介绍一款免费的在线交流工具呢?

A: 当然可以，其实有很多的免费在线交流工具可供你选择。对于新手，我一般推荐博客。

B: 嗯，虽然我以前听说过它，但不知道它是一种什么样的工具。

A: 博客是一种既免费、又快速、又好用的发布工具。

B: 我该怎么用博客和朋友们联系呢?

A: 首先你需要设置个人的登录信息来建立一个账号，之后才能开始使用。

B: 建立账号之后就可以发布消息了吗?

A: 还不可以，你还需要认真考虑给你的博客取一个名字并选择博客的模板。

B: 我明白了。非常感谢。

A: Can I help you?

B: I want to communicate with friends on the Internet. Could you introduce a free online communication tool to me?

A: Sure, actually there are many free tools for online communication. For a newcomer, I recommend blog.

B: Yeah, although I had heard it before, I don't know what kind of tool it is.

A: Blog is a free, fast, and easy-to-use publishing tool.

B: What should I do to contact with friends through blog?

A: First you need personal login information such as username and password to create a blog account to get started.

B: Can I publish information after creating an account?

A: Not really, you should give a lot of thought to choosing a name and a template for your blog.

B: I see. Thank you very much.

4. Oral Practice

Emma要用博客发表第一篇博文了，请告诉她具体的步骤：1. 登录后单击进入发表博文页面；2. 在标题栏处输入博文的标题；3. 编写博文并排版后就可以发表了。

Key Words

博客 blog	发表 publish	博文 post	具体的 concrete
步骤 step	登录 log-in	单击 click	标题栏 title field
编写 write	排版 format		

Part 2 Reading

Text

Tools for Online Communication

There are a variety of tools that are used in an online setting to communicate and collaborate.[1] Because many of these tools are used in real-time settings, rely on different media (visual, auditory, etc.) and can have complex interfaces. They face a number of accessibility barriers.

Q1 Briefly describe what are blogs.

The following resource introduces a few of the most popular technologies used to communicate on and over the Web.

Blogs

What are blogs? Blogs look like Web sites with dated contributions regularly added. They are a form of self publishing where anyone can communicate with the world using various media.[2] Text-based blogs with some illustrations presently predominate. There is no external editor. Publishing is immediate. You critically evaluate to determine the reliability.

Q2 Why do you use blogs?

Why use blogs? Blogs are a simple way for anyone to publish on the Internet. Six of the most common reasons for people to blog include:

- to improve customer/client relations;
- to market a product or range of products;
- to show knowledge and earn credibility for the author;
- to cater for the needs of a niche market[3];
- for self publishing;
- for the sharing of opinions.

Chat

Q3 What is chat?

What is chat? Chat is one of the original communication tools. It was the only tool that provided Internet users with almost instant communication as it is a synchronous tool. Each person uses a computer keyboard to type their ideas and responses to one other person or a group. They read the reply on their monitor. The speed of the reply is dependent on the Internet speed and also on the speed of the typing. The message at the top of the screen is the latest message written. Old messages move down the window and eventually disappear.

Why use chat? People use chat for many reasons:

- it is instantaneous communication with one or more;
- participants can both contribute and "listen" at the same time;
- chatting can occur while other forms of media are being used, e. g. asking questions during a seminar;
- can be used for an online meeting;
- it bears a similarity to SMS including the use of acronyms;
- if it is part of a professional learning activity, a transcript can be printed to use as evidence of the learning.

Q4 *Why do you use chat?*

Email

What is email? Email is electronic mail. It is the sending of "letters" or messages using email accounts and the networks of the Internet. There are two different types of email: email using a desktop application and Web-based email.

Q5 *What is email?*

Why use email? Email was the first universally available online communication tool. Everyone could have an email address even if they did not have Internet access at home. With email you can have your own inbox and address book—a place where you can privately manage your correspondence with others.

Q6 *Why do you use email?*

One of the foremost reasons for businesses using email is the paper trail that is created. This trail can be used to verify that communication has occurred and provide evidence of:

- quotes for products or services;
- instructions provided;
- questions asked and answers given;
- idea development;
- decisions made.

Email is also a very efficient method of sharing files with others by the use of attachments.

Forums

What are forums or discussions? Forums are one of the original online communication tools. They have developed in capability and are still used particularly with online learning communities. They are pages on the Internet where people can post messages and possibly respond to messages posted by others. They are known as an asynchronous communication.

Q7 *What are forums?*

Forums are known as: discussions, message or bulletin board and groups.

Why use forums? Forums are used for many reasons. Some of these are:

Q8 *Why do you use forums?*

- for exchange of information and opinions;
- self help—using specific topic forums to seek answers from the membership;
- sharing expertise and answering questions posted on the forum;
- for collaborating and creating knowledge;
- discussing ideas;
- providing a way for geographically disparate people to interact;
- building an online community.

Microblogs

What are microblogs? Microblogs are the current new online communication tool. A microblogger uses very small posts (usually 140 characters—about the same as a mobile text message) to communicate messages over the Internet to anyone interested in reading them. Microblogs are added by using: any Internet connected computer or a mobile phone.

Q9 *What are microblogs?*

Why use microblogs? It can help people find the information they need more easily, or to get some information about the world they didn't know before. It's convenient—people don't need to use too much time to express their feelings. They can write at most 140 words, and then send it through computers or mobile phones onto the microblogs. As soon as they send it, their followers will know. They can use it anytime, anywhere.

Q10 *Why do you use microblogs?*

Q11 *What is video conferencing?*

Video Conferencing

What is video conferencing? This is one of the newer communication tools. Its popularity has grown with increased speeds of the Internet becoming available. Each person in the video conference call has access to a video camera and microphone that can transmit both moving images and sound across the internet.

Why use video conferencing? The main reason for using the video conference tool is to enable everyone to see the people they are talking with. In a one-on-one conference the other person will always be on the screen. In the case of larger conferences the person speaking will be the one seen on the screen.

Q12 Why do you use video conferencing?

VoIP[4]

What is VoIP? VoIP (voice over the Internet protocol) is one of the newest communication tools. It allows Internet users to make telephone calls. Modems allow computers to use telephone lines to communicate data. VoIP allows users to talk to others using the computer. VoIP uses the network structure of the Internet and transmits voice. It is becoming more popular as it provides low cost telephony. Voice can be transmitted using a headset and microphone, the inbuilt microphone/speakers and a telephone designed for VoIP.

New Words

real-time ['riəltaim]	*a*. [计算机] 实时的 e. g. New cars often include navigation systems, some with real-time traffic updates so that drivers can avoid jams. 新车通常装有导航系统，有些还能实时更新交通信息，这样司机就可以避开交通堵塞。
visual ['vizjuəl]	*a*. 视力的，视觉的 e. g. It turns out that most of the elements of intuitive graphical interfaces are actually visual idioms. 事实证明，在直觉的图形界面上，大多数元素都采用了视觉习惯用法。

auditory [ˈɔːditəri]	*a*. 听的，听觉的 e. g. Reading things aloud to yourself combines the visual and auditory learning styles. 大声朗读结合了视觉和听觉学习方式。
interface [ˈintəfeis]	*n*. [计算机] 界面，接口，连接，联系装置 e. g. In professional applications, we will want to customize our logic and user interfaces. 在专门应用程序中，我们都需要定制自己的逻辑和用户界面。
accessibility [əkˌsesəˈbiləti]	*n*. 可访问性，可接近性，可利用率 e. g. Accessibility is an important part of the HTML standard. 可访问性是 HTML 标准的重要组成部分。
predominate [priˈdɔmineit]	*vi*. 居支配地位，统治，占主导地位 e. g. The views of the left wing have tended to predominate within the party. 左翼的观点趋向于在该党内部占支配地位。
external [ikˈstəːnəl]	*a*. 外的，外部的，外面的，外侧的 e. g. As consumer spending has slumped, the external imbalances have shrivelled. 随着消费者支出下降，外部失衡已经消退。
synchronous [ˈsiŋkrənəs]	*a*. 同时发生的，并发的，同时的 e. g. Counters can be generally classified as either serial or synchronous. 计数器通常可分为串行和同步两类。
acronym [ˈækrəunim]	*n*. 首字母组合词，首字母缩写 e. g. SCSI is the acronym for "small computer system interface". SCSI 是"小型计算机系统接口"的首字母缩写。
electronic [ˌilekˈtrɔnik]	*a*. 电子的，电子学的 e. g. He used an electronic device to measure and record the eye movements. 他用一种电子仪器来测量和记录眼球运动。
send [send]	*vt*. 发送，寄，派遣，使进入，发射 e. g. Specify fax device settings to enable your computer to send and receive faxes. 指定传真设备设置，使计算机可以发送和接收传真。
accounts [əˈkaunt]	*n*. 账目，账号 e. g. The computer account for the server was not disabled. 该服务器上的计算机账号并未禁用。
available [əˈveiləbl]	*a*. 可得到的，可获得的，可达到的 e. g. Cannot open switch port. Scanning will not be available. 无法打开切换端口。无法使用扫描。

capability [ˌkeipə'biləti]	*n*. 能力，才能，本领，技能 e. g. He is a careful and capable person. I trust his capability. 他是一个细致而能干的人，我相信他的能力。
character ['kærəktə]	*n*. [计算机] 字符 e. g. Places selected character in the Characters to Copy box. 请选中"需复制的字符"框中的字符。
transmit [trænz'mit, træns-]	*vt*. 传导，传递，传达，输送 e. g. He has transmitted the report to us. 他已经把报告传送给我们了。
inbuilt ['inˌbilt]	*a*. 内置的，内藏的，嵌入的（=built-in） e. g. At the click of a button, the centre of the SmarTable rises to reveal the inbuilt fridge. 按下按钮，一台嵌入式的冰箱便会从"智能桌"的中部冒出来。

Key Terms

accessibility barrier 可利用率障碍
text-based 基于文本的
niche market 利基市场，缝隙市场
instant communication 即时通信
seminar 研讨会
SMS (short messaging service) 短信服务
Web-based 基于网络的
paper trail 书面记录
attachment 附件
asynchronous communication 异步通信，非同步通信
bulletin board 布告牌，电子公告栏
one-on-one 面对面的，一对一的
headset 耳机，头戴式受话器

Useful Expressions

a variety of 种种，各种各样的

e. g. Newspapers offer information, and they also cover a variety of topics. 报纸涵盖了大量主题，能提供很多信息。

rely on 依靠，依赖

e. g. If we rely on these excessively, they can make our body functions degenerate gradually. 如果我们过分依赖这些，它们会让我们的身体机能逐渐退化。

cater for 满足（需要），迎合

e. g. we aren't able to cater for your particular needs. 我们不能满足你的特殊需要。

Notes

1. There are a variety of tools that are used in an online setting to communicate and

collaborate.

that 引导定语从句修饰 tool，that 在从句中作主语。

2. They are a form of self publishing where anyone can communicate with the world using various media.

where 引导定语从句修饰 self publishing，where 在从句中作状语。

3. niche market

利基市场，指那些被大市场中的统治者或有绝对优势的企业忽略的某些高度专业化的细分市场。中小企业可以选定一个很小的产品或服务领域，集中力量进入并成为领先者，从当地市场到全国再到全球，同时建立各种壁垒，逐渐形成持久的竞争优势。

4. VoIP (voice over Internet protocol)

互联网协议电话，简而言之就是将模拟声音信号（voice）数字化，以数据封包（data packet）的形式在 IP 数据网络（IP network）上做实时传递。VoIP 最大的优势是能广泛地采用 Internet 和全球 IP 互连的环境，提供比传统业务更多、更好的服务。VoIP 可以在 IP 网络上便宜地传送语音、传真、视频和数据等业务，如统一消息、虚拟电话、虚拟语音/传真邮箱、查号业务、Internet 呼叫中心、Internet 呼叫管理、电视会议、电子商务、传真存储转发和各种信息的存储转发等。

Exercises

1. Match each of the following terms with its Chinese equivalent.

1) text-based　　a. 书面记录
2) instant communication　　b. 布告牌，电子公告栏
3) Web-based　　c. 基于文本的
4) paper trail　　d. 即时通信
5) niche market　　e. 附件
6) asynchronous communication　　f. 基于网络的
7) bulletin board　　g. 异步通信，非同步通信
8) attachment　　h. 虚拟会议
9) headset　　i. 利基市场
10) virtual conference　　j. 耳机

2. Recognize the following abbreviations by matching them with their corresponding full names and translate them into Chinese.

1) VoIP ______________　　a. short messaging service
2) SMS ______________　　b. really simple syndication
3) RSS ______________　　c. voice over the Internet protocol

4) BBS ________________ d. bulletin board system

3. Complete the following sentences by translating the Chinese in the brackets.

1) ________________ (实时动画) requires display hardware capable of displaying a sequence with tens of different images every second.

2) We use ________________ (分屏模式) to show the text being worked on and another text from memory for comparison.

3) The stack is a structure of ________________ (抽象数据类型), it can store any type of data from an integer to an address.

4) The new graphics adapter is capable of displaying ________________ (更高分辨率的图像).

5) All databases can ________________ (输入和输出) a comma-delimited file format.

4. Choose the best answer for each blank.

1) Computer ________ is a complex consisting of two or more connected computing units, it is used for the purpose of data communication and resource sharing. (2006年下半年程序员考试试题)

A. storage B. device C. network D. processor

2) With ________ you can communicate in real time with people all around the world. (2009年下半年程序员考试试题)

A. E-mail B. WWW C. FTP D. Web chat

3) Office ________ is the application of the computer and communications technology to improve the productivity of office workers. (2009年下半年程序员考试试题)

A. Tool B. Automation C. Device D. FAX

4) The ________ in e-mail messages has affected almost every computer around the world and has caused the damage of up to US$ 1 billion in North America. (2007年上半年程序员考试试题)

A. illness B. virus C. weakness D. attachment

5) ________: A location where data can be temporarily stored. (2006年下半年程序员考试试题)

A. Area B. Disk C. Buffer D. File

Supplementary Reading

Comparing Wikis with Other Online Communication Tools

Wikis are a collaborative writing space on the Internet. They can be public or private. Each wiki is generally created on a specific topic. The information available to users grows with the contributions from the wiki users. Here's how a wiki differs from other forms of

Internet collaborative tools, such as e-mail, blogs, bulletin boards, forums, content management systems, and Web publishing systems.

- **Wikis are not e-mail**. Individual e-mails share some wiki properties—they are easy to create, they can be quickly formatted, and almost anyone can create an e-mail. And, e-mail can also be used for one-to-one or multiple communication by sending mail to many people or by using mailing lists. However, e-mail lacks a central place where everyone can work at once. And e-mail also doesn't allow many authors to work on the same page or for pages to be linked. E-mails are also usually short whereas wiki pages can be as long as needed.
- **Wikis are not blogs**. A blog is a set of pages on which short entries are posted, usually appearing in a list with the most recent entries on top. Comments can appear attached to each posting. RSS (really simple syndication, a format for live online data feeds) feeds allow people to be notified when new blog content appears. (Note that RSS feeds can apply to any sort of content, but they seem to be wildly popular with blogs.)

Wiki pages can be made to look like blog pages, but they don't come out of the box with all the pages needed to automatically write and publish blog entries. Blogs are usually focused on one-to-many communication, but wikis are more oriented to many-to-many communication about shared content.

- **Wikis are not bulletin boards or forums**. Bulletin boards (sometimes called forums) are Web pages where you can ask a question, make a comment, or put forth a proposition to which others can respond. The list of comments about a topic appears in a long list of entries, which sometimes branches into subtopics.

Wiki pages can be used like bulletin boards in a style called thread mode, in which new comments are added to the bottom of a wiki page, but this is a style (not a structure) that is enforced by the wiki. In bulletin boards, the structure of the pages and the communication are always the same and cannot be changed by the people using the board.

- **Wikis are not content management or Web publishing systems**. Content management and Web publishing systems are general purpose engines for creating all sorts of Web sites. Like wikis, content management systems are toolkits; unlike wikis, though, they aren't governed by the rules set down to define wikis.

Almost any kind of Web site, blog, bulletin board system, and wiki can be built by a content management system. Many content management systems have extensions to allow wikis to be included in the Web sites that are built. Usually, content management systems can only be used by expert programmers, but wikis can be used right away by almost anyone.

New Words and Expressions

wiki 一种多人协作的写作工具。wiki 站点可以有多人（甚至任何访问者）维护，每个人都可以发表自己的意见，或者对共同的主题进行扩展或者探讨。

thread［θred］*n*. 线，线状体，［机械工程］螺纹，［计算机］（公告板上讨论的）话题，题材

RSS *abbr*.（really simple syndication）简易信息聚合，是一种描述和同步网站内容的格式，是目前使用最广泛的资源共享应用，可以视为资源共享模式的延伸。

Reading Comprehension

Read the following statements below and decide if they are true（T）or false（F）according to the passage you have just read.

1）Wikis are public writing space.（　　）

2）Wiki pages can not be made to look like blog pages.（　　）

3）E-mail can be used for one-to-one or multiple communication by sending mail to many people or by using mailing lists.（　　）

4）Wikis are Web publishing systems.（　　）

5）Many content management systems have extensions to allow wikis to be included in the Web sites that are built.（　　）

Part 3 Screen English

改变账号的图像

提示信息	含义
SUMMARY：Change the picture that appears next to your account in the Windows XP welcome screen. Whenever you log into your Windows XP account，a picture is normally displayed next to the account names. If you'd like to personalize your Windows XP account with a picture of your choice，do the following：SUMMARY：Change the picture that appears next to your account in the Windows XP welcome screen.	概要：改变在 Windows XP 欢迎屏幕上显示在账号后边的图像。 一旦你登录 Windows XP 账号，通常在紧跟账号名称的后边会显示一个图像。如果你想另选图像以个性化你的 Windows XP 账号，可按如下步骤操作： 1）单击“开始”菜单。 2）选择“控制面板”。 3）出现“控制面板”后，选择“用户账号”。 4）出现“用户账号”窗口后，在“或者

（续）

提示信息	含义
Whenever you log into your Windows XP account, a picture is normally displayed next to the account names. If you'd like to personalize your Windows XP account with a picture of your choice, do the following: 1) Click the Start menu. 2) Select Control Panel. 3) When the Control Panel appears, select User Accounts. 4) When the User Accounts window appears, underneath Or Pick An Account to Change, click your account name. 5) Click Change My Picture. 6) Choose one of the available pictures and click Change Picture. Or, click Browse for More Pictures and use a picture of your choice. 7) Close the windows and dialog boxes when done.	选择需要改变的账号”下面，单击账号名。 5）单击“改变我的图像”。 6）选择一个可用图像并单击“改变图像”。或者，单击“浏览更多的图像”并使用你所选择的图像。 7）完成后关闭窗口和对话框。

Quotation

There's a phrase in Buddhism, "Beginner's mind". It's wonderful to have a beginner's mind.

——Steve Jobs

佛教中有一句话叫："初学者的心态"。拥有初学者的心态是件了不起的事情。

Remembering that you are going to die is the best way I know to avoid the trap of thinking you have something to lose.

——Steve Jobs

谨记自己总会死去，是让自己避免陷入“人生有所失”思考的最佳方法。

Key to Exercises

Listening & Speaking

1. 1) telephone　2) Internet　3) internationally　4) act　5) share
   6) talk　7) type　8) available　9) tool　10) visual

2. 1) A　2) A　3) B　4) A　5) B

Exercises

1. 1) c 2) d 3) f 4) a 5) i 6) g 7) b 8) e 9) j 10) h
2. 1) c 互联网协议电话 2) a 短信服务
   3) b 简易信息聚合 4) d 电子公告牌系统/简报板系统
3. 1) Real-time animation 2) split screen mode
   3) abstract data types 4) higher resolution graphics
   5) import and export
4. 1) C 2) D 3) B 4) B 5) C

Comprehension

1) F 2) F 3) T 4) F 5) T

Electronic Commerce

Learning Objectives

After completing this unit, you will be able to:

1. Be familiar with words and expressions involved in e-commerce;
2. Describe three basic types of e-commerce: B2B, B2C and C2C;
3. Buy and sell goods online.

Part 1 Listening & Speaking

1. Listen to the following passage and fill in the blanks with the words in the box.

CEO	founder	Internet-based	Internet	keyboard	home pages
e-commerce		online	small and medium-sized		underdog

Host: Hello and welcome to Talk Asia, a special edition coming to you from Beijing. Our guest today is the 1) ________ and 2) ________ of Chinese 3) ________ giant Alibaba. com—Jack Ma. Born in Hangzhou in 1964, Ma first made a name for himself with his idea to start an 4) ________ directory called "China Pages", widely regarded to be China's first 5) ________ company. In 1999, Ma founded "Alibaba. com", which currently operates the world's largest online marketplace, serving 6) ________ enterprises by connecting global buyers with sellers in China. In 2005, Alibaba. com added to its international profile by acquiring and operating Yahoo China. Ma himself has impressed many with his sharp entrepreneurial sense and work ethic. Mr. Ma, thank you for joining us. It's good to see you. You have for years carried the image as the 7) ________. But this particular underdog has done very well for him, hasn't he?

Ma: Yes, we've been sort of very lucky.

Host: You emerged from virtually nowhere, right? And you started this first Chinese 8) ________ company. Where did this idea come from?

Ma: In 1995, I went to Seattle and My friend said "Jack, this is the Internet." I have never heard about the Internet and never used the computer before that day. My first time in life, I touched on the 9) ________ and I find well—this is something I believe; it is something that is going to change the world and change China.

Host: What about China Pages?

Ma: China Pages, it was just, well, we were making 10) ________ for Chinese companies and listing it in the USA to let the world know about Chinese companies. That's the earliest e-commerce, and that's probably the first Internet company in China.

Host: But only with, I read, 2,000 dollars?

Ma: Yup, I borrowed 2,000 dollars from my relatives—my brother-in-law and my parents. So we started the company.

2. Choose the proper words or expressions.

Alibaba. com is the world's leading online B2B 1) ________ connecting millions of 2) ________ around the world every day. With 6.9 million international members from over 240 countries and regions, Alibaba. com is the best choice for anyone involved in international trade! Buyers have 3) ________ to 4.2 million suppliers, so our powerful keyword search makes finding the products you need simple. Our 4) ________ always rank above suppliers who display products for free, making it easy to access our best suppliers fast. If you want to sell your products or services online, Alibaba. com is also a cost-effective marketing channel. As a free member, you will be given a free company website to showcase your products or services to millions of buyers around the world. With our 5) ________ package, you'll be the choice of over 85% of buyers on Alibaba. com who prefer to do business with verified members only. Please read our detailed "How to Buy and Sell" guide and let us help you get the most out of Alibaba. com.

1) A. trade platform　　B. trade frame
2) A. buyers and sellers　　B. buyers and suppliers
3) A. access　　B. assess
4) A. verified suppliers　　B. very suppliers
5) A. Precious Membership　　B. Premium Membership

3. Please read the conversation below and learn how to raise a question and how to solve each problem.

货物 goods	电子手段 electronic means
交易 transaction	商业机会 commercial opportunity
成本 cost	提升服务质量 better the quality of services
黄金供货商 Gold Supplier	订单 order

A：你好，马克，我需要你的建议。
B：你说吧，琳达。
A：我们现在有很多货物积压卖不出去，该怎么办呀？
B：别担心。事实上我进入 B2B 世界之前也遇到过类似的问题。
A：什么是 B2B？
B：它指的是企业之间利用电子手段进行交易。B2B 能创造更多的商业机会，降低成本，还能提升服务质量。
A：听起来不错。你能讲详细点吗？
B：好的，给你推荐一个著名的 B2B 网站：Alibaba. com. 自从我们公司成为 Alibaba. com

的黄金供货商，我们的订单比以前多多了。真的很棒。

A：太好了！我们坐下来你给我演示一下 B2B 是怎么运作的吧。

B：好的。

A: Hello, Mark, I need your suggestion.

B: Go ahead, Linda.

A: We have lots of goods in stock now but we cannot sell them out. What can we do?

B: Don't worry. Actually, I met the same problem before I entered the B2B world.

A: What is B2B?

B: It refers to transactions conducted by electronic means between businesses. It can create more commercial opportunities, reduce costs, and better the quality of services.

A: Sounds good. Tell me more about it.

B: OK, let me recommend to you a famous B2B website: Alibaba.com. Since our company became a Gold Supplier in Alibaba.com, we have got much more orders than before. It's amazing.

A: Great! Let's take a seat and show me how exactly B2B works.

B: OK.

4. Oral Practice

假设你的一位外国朋友想在淘宝上买东西，但他不知道如何操作，于是向你求助。请利用以下关键词帮助他。

Key Words

主页 home page	搜索栏 search bar	购物车 shopping cart
搜索 search	输入 input	类别 category
比较 compare		

Part 2 Reading

Text

Electronic Commerce

Electronic commerce[1], also known as e-commerce, is the buying and selling of goods over the Internet. Have you ever bought anything over the Internet? If you have not, there is a very good chance that you will have within the next year or two. Shopping on the Internet is growing rapidly and there seems to be no end in sight.

The underlying reason for the rapid growth in e-commerce is that it provides incentives for both buyers and sellers. From the buyer's perspective, goods and services can be purchased at any time of day or night. Traditional commerce is typically limited to standard business hours when the seller is open. Additionally, buyers no longer have to physically travel to the seller's location. For example, busy parents with small children do not need to coordinate their separate schedules or to arrange for a baby sitter whenever they want to visit the mall. From the seller's perspective, the costs associated with owning and operating a retail outlet can be eliminated. For example, a music store entirely on the Web without an actual physical store and without a large sales staff. Another advantage is reduced inventory. Traditional stores maintain an inventory of goods in their stores and periodically replenish this inventory from warehouses. With e-commerce, there is no in-store inventory and products are shipped directly from warehouses.

Q1 *What is electronic commerce?*

Q2 *What is the underlying reason for the rapid growth in e-commerce?*

While there are numerous advantages to e-commerce, there are disadvantages as well. Some of these disadvantages include the inability to provide immediate delivery of goods, the inability to "try on" prospective purchases, and questions relating to the security of online payments. Although these issues are being addressed, very few observers suggest that e-commerce will replace bricks-and-mortar businesses entirely. It is clear that both will coexist and that e-commerce will continue to grow.

Just like any other type of commerce, electronic commerce involves two parties: businesses and consumers. There are three basic types of electronic commerce:

Q3 *What are the three basic types of e-commerce?*

B2C (business to consumer)[2] involves the sale of a product or service to the general public or end users. Oftentimes this

Q4 *What is B2C?*

arrangement eliminates the wholesaler by allowing manufacturers to sell directly to customers. Other times, existing retail stores use B2C e-commerce to create a presence on the Web as another way to reach customers.

C2C (consumer to consumer)[3] involves individuals selling to individuals. This often takes the form of an electronic version of the classified ads or an auction.

Q5 *What is C2C?*

B2B (business to business)[4] involves the sale of a product or service from one business to another. This is typically a manufacturer-supplier relationship. For example, a furniture manufacturer requires raw materials such as wood, paint, and varnish.

Q6 *What is B2B?*

New Words

underlying [ˌʌndəˈlaiiŋ]	*a*. 潜在的 e.g. To stop a problem you have to understand its underlying causes. 要解决问题，你得了解其潜在原因。
incentive [inˈsentiv]	*n*. 刺激，鼓励 e.g. Awards provide an incentive for young people to improve their skills. 奖赏鼓励年轻人努力提高自己的技能。
perspective [pəˈspektiv]	*n*. 视角，观点 e.g. His father's death gave him a whole new perspective on life. 父亲的死使他对生活有了全新的看法。
purchase [ˈpəːtʃəs]	*vt*. 买，购买 *n*. 购买，购买的物品 e.g. You can purchase insurance online. 你可以在网上买保险。
physically [ˈfizikli]	*ad*. 身体上地，实际上 e.g. She is young and physically fit. 她年轻而健康。
coordinate [kəuˈɔːdineit]	*vt*. 调节，协调 e.g. The cooker is green, to coordinate with the kitchen. 厨灶是绿色的，为了和厨房的颜色协调。
baby sitter [ˈbeibiˌsitə(r)]	*n*. 临时保姆
inventory [ˈinvəntəri]	*n*. 存货 e.g. In our warehouse you'll find a large inventory of new bikes. 在我们的仓库里，你会发现大量的新自行车存货。

periodically [ˌpiəriˈɔdikli]	ad. 周期性地，定期地
replenish [riˈpleniʃ]	vt. 补充 e. g. The music will replenish my weary soul. 音乐使我疲惫的精神充满活力。
warehouse [ˈweəhaus]	n. 仓库，货栈 e. g. Be sure to lock the warehouse when you leave. 你离开仓库时，请一定把它锁上。
in-store [inˈstɔː]	a. 库存的
advantage [ədˈvɑːntidʒ]	n. 优点，有利条件（因素），好处 e. g. This school has many advantages. 这个学校有很多优点。
disadvantage [ˌdisədˈvɑːntidʒ]	n. 不利（条件） e. g. Do your disadvantayes over weight the advantages? 弊多于利吗?
delivery [diˈlivəri]	n. 发送，交货 e. g. Please advise the date of delivery. 请告知发货时间。
security [siˈkjuəriti]	n. 安全 e. g. The exact locations are being kept secret for reasons of security. 确切的地点因为安全原因要保密。
payment [ˈpeimənt]	n. 支付的款项，支付，付款 e. g. We are saving for a down payment on a house. 我们正攒钱支付买房的首付金。
bricks-and-mortar [ˈbriksˌəndˈmɔːtə]	n. 传统的实体企业 e. g. Bricks-and-mortar shops struggle to win customers back from virtual ones. 实体商场从网络商家手中极力挽回消费者。
eliminate [iˈlimineit]	vt. 排除，消除 e. g. America wants to eliminate tariffs on items such as electronics. 美国打算取消电子产品等的关税。
wholesaler [ˈhəulˌseilə]	n. 批发商 e. g. We're the largest furniture wholesaler in China. 我们是中国最大的家具批发商。
manufacturer [ˌmænjuˈfæktʃərə]	n. 制造商，制造厂 e. g. The manufacturers in some countries dumped their surplus commodities abroad. 一些国家的制造商向国外倾销过剩产品。

retail ['riːteil]	*n*. 零售 e. g. The retail dealer buys at wholesale and sells at retail. 零售商批发购进货物，以零售价卖出。
customer ['kʌstəmə]	*n*. 顾客，主顾 e. g. To save customers' time, they began a delivery service. 为了节省顾客的时间，他们实行送货上门。
individual [ˌindi'vidʒuəl]	*n*. 个人（体） e. g. Tourism is up, jobs are up, individual income is up. 旅游火了，工作机会多了，个人收入涨了。
auction ['ɔːkʃən]	*n*. & *vt*. 拍卖 e. g. The house will be sold by auction. 这栋房子将要拍卖。
supplier [sə'plaiə]	*n*. 供应者，供应商
varnish ['vaːniʃ]	*n*. 清漆

Key Terms

electronic commerce 电子商务
B2C 企业对消费者
C2C 消费者对消费者
B2B 企业对企业

Useful Expressions

in sight 看得见，被看到，在即，在望

e. g. An island is coming in sight. 一个小岛渐入视野。

provide for 供给，为……作准备

e. g. They worked hard to provide for their large family. 他们努力工作以供养一大家子人。

Notes

1. electronic commerce

电子商务是利用计算机技术、网络技术和远程通信技术，实现整个商务（买卖）过程中的电子化、数字化和网络化。它是借由网络，通过网上琳琅满目的商品信息、完善的物流配送系统和方便安全的资金结算系统进行交易。

2. B2C (business to customer)

企业对消费者的电子商务模式。这种形式的电子商务一般以网络零售业为主，主要借

助于 Internet 开展在线销售活动。

3. C2C (consumer to consumer/customer to customer)

消费者对消费者的电子商务模式。比如一个消费者有一台旧计算机，通过网络进行交易，把它出售给另外一个消费者，此种交易类型就称为 C2C。

4. B2B (business to business)

企业对企业的电子商务模式，也泛指企业间的市场活动，不局限于最终交易对象的认定。另外，B2B 也指企业间定义业务形态的方式。B2B 着重于企业间网络的建立、供应链体系的稳固。目前中国有名的 B2B 网站有阿里巴巴、慧聪、中国制造网等。

Exercises

1. Match each of the following terms with its Chinese equivalent.

1) e-commerce　　a. 黄金供货商
2) shopping cart　　b. 在线订货
3) Gold Supplier　　c. 电子商务
4) order　　d. 电子支票
5) e-check　　e. 订单
6) issuing bank　　f. 配送中心
7) online ordering　　g. 数字证书
8) distribution center　　h. 购物车
9) digital certificate　　i. 数字签名
10) digital signature　　j. (外贸) 开证银行

2. Recognize the following abbreviations by matching them with their corresponding full names and translate them into Chinese.

1) B2B ________　　a. electronic funds transfer
2) B2C ________　　b. commerce service provider
3) C2C ________　　c. point of sale
4) POS ________　　d. business to consumer
5) PPS ________　　e. consumer to consumer
6) EDI ________　　f. business to business
7) SMEs ________　　g. payment processing service
8) EC ________　　h. electronic commerce
9) CSP ________　　i. small and medium-sized enterprises
10) EFT ________　　j. electronic data interchange

3. Complete the following sentences by translating the Chinese in the brackets.

1) With e-commerce, there is no in-store inventory and ________ (产品可直接从仓库发出).

2) Electronic commerce, also known as e-commerce, is ________ (在因特网上买卖商品).

3) The underlying reason for ________ (电子商务快速发展) is that it provides incentives for both buyers and sellers.

4) While there are numerous advantages to e-commerce, ________ (同时也存在着缺陷).

5) B2B (business to business) involves ________ (产品或服务由一家企业出售给另一家).

4. Choose the best answer for each blank. (2001年网络程序员考试上午试题)

The major problem with E-mail is that it is 1) ________ easy to use that people can become 2) ________ with messages 3) ________ they can possibly answer in a day. In addition, mail boxes require some management to 4) ________ messages or archive those that might be required later. Senders don't always know about your E-mail backlog and often send 5) ________ messages.

1) A. too B. so C. very D. much

2) A. full B. lost C. inundated D. filled

3) A. more than B. than C. that D. which

4) A. manage B. save C. backup D. dispose of

5) A. too many B. redundant C. long D. trivial

Supplementary Reading

Taobao

It's reported that Taobao, Alibaba's online consumer website, expects its transaction volume to exceed 100 billion yuan (14.4 billion U.S. dollars) in 2008, a 130 percent increase from the previous year.

There's no denying that Taobao has made spectacular development in terms of on-line shopping. Before specifying on its business model, I would give a briefing on Taobao. Taobao is one of seven unlisted arms under the Hong Kong-listed Alibaba.com which was founded in 1999 by Jack Ma in Hangzhou. Its shares grew 122 percent on their trading debut in November. Since its founding in 2003, it has developed into the top domestic on-line shopping website, accounting for nearly 70% market shares in China.

As an on-line C2C (consumer to consumer) platform for personal transaction, Taobao website is becoming the top choice for on-line business.

On Chinese C2C market, eBay is Taobao's biggest rival. In March 2002, eBay invested 30 million dollars to establish eBay (China) or Yiqu. In July 2003, eBay put another additional capital injection of 150 million dollars into eBay (China). Up to now, such firms as Motorola, Nokia, Haier, Lenovo, Great Wall, Apple, Adidas have set up their on-line shops on Taobao website.

Market, payment, creditability and searching capability are four elements of e-commerce. Now its consumer arms include the online payment units Alipay and Yahoo China. Aliplay is available to over 100,000 on-line shops and becomes the most popular and safest on-line payment tool.

The business model for Taobao aims to cut business cost, resulting in the win-win situation between manufacturers and consumers. Consumers can often get a great bargain or find a hard-to-find object. At present, the major consumer group of Taobao is aged between 18 and 35. They pursue fashion and have relatively higher consuming capability.

When it comes to the most popular on-line commodities, people tend to purchase inexpensive products, say, clothes, books, IT products and so on. However, no matter how popular it becomes, on-line shopping is not always satisfactory. The development of e-commerce also needs improvement. Taobao is no exception. One of the great problems of shopping on line is the delay and uncertainness of delivery time. In reality, however, not many sellers can deliver goods within 24 hours. Therefore, in their efforts to publicize the advantages of shopping on line, on-line sellers should take timeliness into consideration. In addition, delivery option is another key issue of concern. Now regular mail is by no means the top choice for shoppers in terms of timeliness. So, sellers should choose reliable express companies, which is also an effective way to expand on-line shopping market.

All in all, like other e-commerce websites, it's better to listen to consumers than just developing registered users. Providing consumers with tangible convenience can better realize its business value.

New Words and Expressions

spectacular [spek'tækjulə] *a*. 壮观的，令人惊叹的

briefing ['bri:fiŋ] *n*. 简要情况，简介

debut ['deibju:] *n*. 首次演出，初次露面

rival [ˈraivəl] *n.* 竞争对手

timeliness [ˈtaimlinis] *n.* 及时，适时

in terms of 就……而言，在……方面

Alipay 支付宝

Reading Comprehension

Read the following statements below and decide if they are true (T) or false (F) according to the passage you have just read.

1) As an on-line C2C (consumer to consumer) platform for personal transaction, eBay website is becoming the top choice for on-line business. (　　)
2) On Chinese C2C market, eBay is Taobao's biggest rival. (　　)
3) Market, payment, creditability and searching capability are four elements of e-commerce. (　　)
4) One of the great problems of shopping on line is the delay and uncertainness of delivery time. (　　)
5) Sellers should choose reliable express companies, which is also an effective way to expand on-line shopping market. (　　)

Part 3 Screen English

常见杀毒软件操作提示

提示信息	含　义
Scanning for known viruses.	正在扫描已知病毒。
This will affect more than one file, continue?	该操作会影响多个文件，是否继续?
Diskette compression successfully completed.	磁盘压缩成功完成。
Since a virus was detected, rebooting is recommended to minimize the possibility of further infection.	检测到病毒，建议重启计算机以免更多文件被感染。
Please contact your McAfee agent, your distribution source, or McAfee Associates directly to obtain the latest version.	为升级到最新版本，请联系你的迈克菲代理、分销商，或者直接与迈克菲总公司联系。

Quotation

eBay might be a shark in the ocean, but Alibaba is a crocodile in the Yangtze.

——Ma Yun

易趣可能是大海里的鲨鱼，但阿里巴巴却是长江里的扬子鳄。

I always tell myself, when doing business in China or everywhere around the world, you have to follow the rules. If you cannot change the law, if you cannot create the law, follow the law.

——Ma Yun

我经常告诉我自己，无论是在中国，还是在世界上任何一个地方做生意，你都必须要遵守法律。如果你不能改变法律，如果你不能建立法律，那么就遵守法律。

Key to Exercises

Listening & Speaking

1. 1) founder 2) CEO 3) e-commerce 4) online 5) Internet
   6) small and medium-sized 7) underdog 8) Internet-based 9) keyboard
   10) home pages
2. 1) A 2) B 3) A 4) A 5) B

Exercises

1. 1) c 2) h 3) a 4) e 5) d 6) j 7) b 8) f 9) g 10) i
2. 1) f 企业对企业 2) d 企业对消费者 3) e 消费者对消费者
   4) c 销售点终端 5) g 支付处理服务系统 6) j 电子数据交互
   7) i 中小企业 8) h 电子商务 9) b 商务服务提供商
   10) a 电子资金划拨
3. 1) products are shipped directly from warehouses
   2) the buying and selling of goods over the Internet
   3) the rapid growth in e-commerce
   4) there are disadvantages as well
   5) the sale of a product or service from one business to another
4. 1) B 2) C 3) A 4) D 5) B

Comprehension

1) F 2) T 3) T 4) T 5) T

Install and Configure Software Programs

Learning Objectives

After completing this unit, you will be able to:

1. Improve your English listening skill and oral English;
2. Know how to install Microsoft Word 2007;
3. Know how to configure Virtual PC for USB printing.

Part 1 Listening & Speaking

1. Listen to the following passage and fill in the blanks with the words in the box.

download	double-click	update	installers
desktop	administrator	website	

Many software 1) ________ and updaters that you 2) ________ from the Internet are disk image (.dmg) files. After you download such a file from a 3) ________ (such as Apple Downloads), an installer window will usually appear.

If you don't see an installer window, simply 4) ________ the downloaded disk image file to mount the disk image on your 5) ________. Double-click the disk volume that appears and then double-click the installer or updater file to start the installation process.

If you see an application instead of an installer, go to "Installing software from a downloaded file that doesn't have an installer." Follow the onscreen instructions to install or 6) ________ the software. You will need to enter an 7) ________ and password to install or update software

2. Choose the proper words or expressions.

A computer is a programmable digital electronic device that—through the use of stored instructions and programs—is able to perform different tasks with data, including compilation, 1) ________, correlation, retrieval, and selection. A computer program is one of two things. It may refer to the coding of software applications done by programmers or developers. Alternatively, it may 2) ________ the 3) ________ of development, 4) ________ computer programs, software applications, computer software, or apps. Even though system software is developed by programmers, it is not characteristically referred to as a computer program: the term is generally used to refer to standalone software or suites that is used in conjunction with system software, but is not required for the computer to run.

There are many different varieties of computer program, and for any task the computer user wishes to perform, there are usually a number of options available. A computer program can be designed for only one operating system, or there may be different iterations for the various major operating systems. If it achieves 5) ________, new versions of a computer program will be released periodically in order to fix bugs, add features, and provide updates to match changes in computer operating systems.

1) A. computation B. computing
2) A. refer to B. be referred to
3) A. finish products B. finished products
4) A. knowing as B. known as
5) A. popularity B. popular

3. Please read the conversation below and learn how to raise a question and how to solve each problem.

网页设计师 Web designer	布局 layout	内容 contents
页面 page	图案 graphics	检索功能 search function
链接 link		

A: 约翰，你好。近来怎么样？还在家里玩计算机，也不给老朋友打个电话？

B: 你好，艾米。你是从哪里打过来的呀？不想过来喝一杯？

A: 现在才十一点，喝酒也早了点。给你打电话是想征求一下你的建议。半小时后我要跟一个网页设计师会面，但问题是我不知道我想要什么。

B: 恐怕没人能告诉你到底你想要什么。当然我可以告诉你设计网站时要记住的基本的 东西。在网站的布局和结构上是有些原则的。

A: 我想我自己可以决定内容，我知道我想在网页上放些什么。

B: 最好是让设计师准备几个不同的计划，这样一来你就可以从中选择你真正喜欢的一个。只是要记住一些基本原则，如：在一个页面上不要放太多内容，不要用太多的图片和图案，用一个网站导航栏帮助浏览者，加一个检索功能，不要放太多的链接。

A: 约翰，我有一个绝好的想法！你何不过来跟我一起和网页设计师谈谈？我说说内容，你谈谈技术规范。

B: 对不起，没门！我边玩计算机边喝啤酒多开心。

A: 快点过来，他就要到了！

A: Hi John, how are you doing these days? Still sitting at home with the computer and not even giving a call to your old friends?

B: Hi Amy, where are you calling from? Don't you want to come over for a beer?

A: It's eleven o'clock. It's too early to start drinking. I have phoned you because I need your advice. I have a meeting with a Web designer in half an hour. The problem is I don't really know what I want.

B: I'm afraid nobody can tell you what you want. But of course I can tell you some basic things to remember when designing a site. There are rules about the layout and structure.

A: I think I can decide about the contents. I know what I want to put on the page.

B: The best thing would be to ask the designer to prepare several different plans so you can choose the one you really like. Just remember some basic rules, like don't load too much information on one page, don't use too many pictures and graphics, use a site map to help the visitor, apply a search function, and don't use too many links.

A: John, I have a brilliant idea! Why don't you come over and talk to this Web designer with me? I could tell him about the content and you could give him the technical specifications.

B: Sorry, no way. I'm quite happy with my PC and a beer.

A: Just come over quickly, because he is arriving in no time!

4. Oral Practice

John 发现 Tony 的浏览器加载网页的速度比自己的快，原来 Tony 修改了 IE 浏览器的代理服务器。于是 Tony 向他介绍如何设置或更改 IE 的代理服务器。

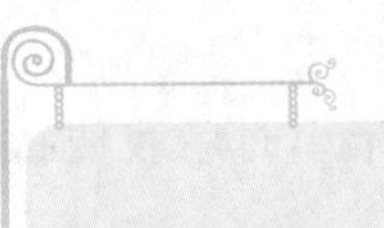

Key Words

代理服务器 proxy server	浏览器 browser	局域网 LAN
链接 connection	高级 advanced	地址框 address box
默认 by default	端口框 port box	

Part 2 Reading

Text

Instructions in Microsoft Word 2007: How to Install Microsoft Word 2007

Before you install Microsoft Word 2007[1] on your computer you must check if your operating system is compatible with the system requirements of the program.

To locate general information regarding your operating system:

Click on the Start button in Microsoft Windows.

Locate My Computer or Computer which is generally located on the right-hand side of the menu. This will open a window displaying folders such as Owner's Documents, DVD/CD-ROM Drive, and Local Disk (C:).

Find the System Tasks menu on the right-hand side of the My Computer window and click on View System Information.

This will open a window displaying general information such as what edition, version, and service pack you currently have installed, as well as how much memory you have available and what processor you are working with.

All Microsoft Word 2007 components have approximately the same system requirements:

Q1 What are the system requirements for Microsoft Word 2007?

A Microsoft Windows XP[2] operating system or later.

A minimum of 256MB of RAM and a recommended 1GB of RAM.

2GB of available disk space on your hard drive.

A Pentium 4 processor is recommended; however, a Pentium III 500MHz processor or higher is required.

A monitor with 16bit colors and Super VGA (1024×768).

A CD-ROM drive and a mouse.

Some additional requirements are necessary to operate specific Microsoft Word 2007 add-ons and/or components:

You will need Windows Internet Explorer 7.0 or higher to access InfoPath 2007[3].

To use speech recognition software, you will need a Pentium III 500MHz processor or higher.

Office OneNote 2007 requires a Tablet PC pen input to capture handwritten notes in digital ink.

The Internet and a webcam are needed to enjoy the full collaboration benefits of Microsoft Exchange Server 2007[4].

How to install Microsoft Word 2007:

Q2 Could you please simply retell the precedure of installing Microsoft Word 2007?

Click Start in Microsoft Windows.

Once the Start menu is visible, click on Control Panel.

From here, click on the Control Panel icon to open the

Control Panel window.

Open Add or Remove Programs. In this window you will see options to add programs from a CD-ROM or as part of a Windows Update.

Click on CD or Floppy.

A window will appear that will walk you through the installation process. First, insert the Microsoft Word 2007 installation disk into your CD-ROM drive and select Next.

After carefully following the instructions in this window, you will be prompted to restart your computer. After doing so you will be notified that a new program has been added. Congratulations! You now have Microsoft Word 2007 on your computer!

New Words

compatible [kəm'pætəbl]	*a*. 兼容的
locate [ləu'keit, 'ləu]	*v*. 定位 e. g. Please locate an electrical fault. 请找出一个电气故障。
edition [i'diʃən]	*n*. 版本
version ['vəːʃən]	*n*. 版本 e. g. This is a new version of Word. 这是 Word 的一个新版本。
approximately [ə'prɔksimitli]	*ad*. 大约
capture ['kæptʃə]	*v*. 采集，捕获 e. g. He captured a baby's smile in a photograph. 他拍摄到一个婴儿的微笑。
collaboration [kəˌlæbə'reiʃən]	*n*. 合作 e. g. She wrote the book in collaboration with her sister. 她与妹妹合写了一本书。
visible ['vizəbl]	*a*. 明显的，看得见的
recommend [ˌrekə'mend]	*v*. 建议，推荐

prompt [prɔmpt]	*v*. *n*. 提示 e. g. The speaker was rather hesitant and had to be prompted occasionally by the chairman. 发言者讲话结结巴巴的，有时还得靠会议主持提示。

Key Terms

folder 文件夹

component 组件

webcam 网络摄像头

add-on 附加软件，附件组件

floppy 软盘

Notes

1. Microsoft Word 2007

Word 是 Microsoft 开发的一个文字处理应用程序，是 Office 系列办公软件中的一个。Word 2007 于 2006 年发布。

2. Microsoft Windows XP

Windows XP 是 Microsoft 发布的一款视窗操作系统，可供个人计算机使用，包括商用及家用的台式机、便携式计算机、媒体中心等，发布于 2001 年 10 月 25 日。

3. InfoPath 2007

InfoPath 是 Microsoft Office 产品的信息收集程序，它使在整个公司内收集和重用信息更加容易。

4. Microsoft Exchange Server 2007

Microsoft Exchange Server 2007 是业界领先的电子邮件、日历和统一信息服务器。作为一套改良的企业统一沟通平台，它支持 x64 平台，创新的 LCR 技术可以极大地降低企业在数据备份和恢复上的投入。

Exercises

1. Match each of the following terms with its Chinese equivalent.

1) add-on　　a. 提示

2) floppy　　b. 建议

3) folder　　c. 网络摄像头

4) component　　d. 组件

5) webcam　　e. 文件夹

6) recommend　　f. 软盘

7) prompt　　g. 附加软件

2. Recognize the following abbreviations by matching them with their corresponding full names and translate them into Chinese.

1) MB ________ a. low level design

2) RAM ________ b. integration testing case

3) VGA ________ c. integration testing plan

4) HLD ________ d. high level design

5) ITP ________ e. video graphics array

6) ITC ________ f. random access memory

7) LLD ________ g. mega byte

3. Complete the following sentences by translating the Chinese in the brackets.

1) Before you install Microsoft Word 2007 on your computer you must check ________ (你的操作系统是否兼容) with the system requirements of the program.

2) Find the System Tasks menu on ________ (右手边) of the My Computer window.

3) Once you have checked ________ (全部的系统要求), you are ready to install Microsoft Word 2007.

4) After carefully following the instructions in this window, ________ (将会提示你重新启动计算机).

4. Choose the best answer for each blank. (2010年下半年程序员考试上午试题(B))

1) ________ means that a program written for one computer system can be compiled and run on another system with little or no modification.

A. Portability B. Reliability C. Availability D. Reusability

2) Data items are added or deleted from the list only at the top of the ________.

A. queue B. stack C. tree D. linear list

3) ________ statement can perform a calculation and store the result in a variable so that it can be used later.

A. Assignment B. Control C. I/O D. Declaration

4) The ________ scheme in a database system is responsible for the detection of failures and for the restoration of the database to a state that existed before the occurrence of the failure.

A. query B. test C. check D. recover

5) Software ________ focuses on three attributes of the program: software architecture, data structure, and procedural detail.

A. analysis B. design C. installation D. upgrade

Supplementary Reading

How to Configure Virtual PC for USB Printing

Virtual PC supports printing to most types of printers, under most guest operating systems. HP inkjet USB printers require that you print using the printer driver supplied with the printer. To use USB printing, you must use a USB cable to connect your printer to your Macintosh computer. Then, you must configure Virtual PC to use the USB connection to your printer with the virtual machine. To do so, follow these steps:

1. Start Virtual PC.
2. Click **Settings** in the **Virtual PC List** dialog box.
3. Click **USB**.
4. Click to select the **Enable USB** check box.
5. Click to select the check box for the USB printer that you want to use.
6. Click **OK**.
7. If your virtual machine is running, restart your virtual machine.

Note When you enable USB in Virtual PC for Mac, the printer uses the virtual USB port of the virtual machine. While the virtual machine is running, the printer is not available to the Macintosh operating system. If the Macintosh operating system tries to use the printer while the Windows-based virtual machine is running, the Macintosh operating system behaves like the printer is physically disconnected and generates an error message.

Typically, the Macintosh computer automatically resumes control over the printer when you close the virtual machine window, or when you quit Virtual PC. Sometimes, the printer may not appear in the Macintosh operating system when Virtual PC releases the printer. If this problem occurs, turn off the physical printer, disconnect its USB cable, turn the printer on, and then reconnect the printer to the host computer.

New Words and Expressions

configure [kən'figə] *v*. 配置
inkjet *a*. 喷墨的
virtual ['vəːtʃuəl] *a*. 虚拟的，有效的
generate ['dʒenəreit] *v*. 使形成，产生
cable ['keibl] *n*. 电缆
resume [ri'zjuːm,-'zuːm] *v*. 重新开始，重新获得

Reading Comprehension

Read the following statements below and decide if they are true (T) or false (F) according to the passage you have just read.

1) Under most guest operating systems, Virtual PC supports printing to only a few types of printers. (　　)

2) The virtual USB port of the virtual machine is used by the printer when users enable USB in Virtual for Mac. (   )

3) The Macintosh operating system behaves like the printer is physically disconnected and generates an error message if the Macintosh operating system tries to use the printer while the Windows-based virtual machine is running. (   )

4) When the virtual machine window is closed, or when Virtual PC is quit, the Macintosh computer can not resumes control over the printer automatically. (   )

Part 3 Screen English

更改密码

提示信息	含　义
1) Click the Start menu. 2) Select Control Panel. 3) When the Control Panel appears, select User Accounts. 4) When the User Accounts window appears, underneath Or Pick An Account to Change, click your account name. 5) Click Change My Password. 6) Follow the on-screen instructions to type in your current and new passwords. You must type your new password twice. You can also enter a hint for Windows XP to display in case you forget your password.	1) 单击"开始"菜单。 2) 选择"控制面板"。 3) 当出现控制面板时，选择"用户账号"。 4) 当出现"用户账号"窗口时，在"或者选择需改变的账号"下面，单击账号名。 5) 单击"更改我的密码"。 6) 按照屏幕指示输入你当前的密码以及新设置的密码。新密码必须被输入两遍。你也可以输入一个当你忘记密码时Windows XP显示的提示信息。

Quotation

Computers do not solve problems, they execute solutions.

——Laurent Gasser

计算机不能解决问题，它们只是执行解决方案。

Key to Exercises

Listening & Speaking

1. 1) installers　2) download　3) website　4) double-click　5) desktop

6) update 7) administrator

2. 1) A 2) A 3) B 4) B 5) A

Exercises

1. 1) g 2) f 3) e 4) d 5) c 6) b 7) a

2. 1) g 兆字节 2) f 随机存取存储器 3) e 视频图形阵列
   4) d 概要设计说明书 5) c 集成测试计划 6) b 集成测试用例
   7) a 详细设计说明书

3. 1) if your operating system is compatible
   2) the right-hand side
   3) all your system requirements
   4) you will be prompted to restart your computer

4. 1) A 2) B 3) A 4) D 5) B

Comprehension

1) F 2) T 3) T 4) F

Computer Security

Learning Objectives

After completing this unit, you will be able to:

1. Know computer viruses and hackers;
2. Learn something about computer security.

Part 1 Listening & Speaking

1. Listen to the following passage and fill in the blanks with the words in the box.

interfere	hidden	attachments	delete
disk	images	software	spread

What is a Computer Virus?

Computer viruses are small 1) ________ programs that are designed to spread from one computer to another and to 2) ________ with computer operation. A virus might corrupt or 3) ________ data on your computer, use your email program to 4) ________ itself to other computers, or even erase everything on your hard 5) ________. Computer viruses are often spread by 6) ________ in email messages or instant messages. Viruses can be disguised as attachments of funny 7) ________, greeting cards, or audio and video files. Computer viruses also spread through downloads on the Internet. They can be 8) ________ in illicit software or other files or programs you might download.

2. Choose the proper words or expressions.

How Can I Prevent Infection by Computer Viruses?

Nothing can guarantee the security of your computer, but there's a lot you can do to help lower the chances that you'll get a virus. It's 1) ________ to keep your 2) ________ software current with the latest updates (usually called definition files) that help the tool 3) ________ and 4) ________ the latest threats. You can continue to improve your computer's security and 5) ________ the possibility of infection by using a firewall, keeping your computer up to date, maintaining a current antivirus software subscription (such as Microsoft Security Essentials), and following a few best practices. 6) ________ no security method is guaranteed, it's important to back up critical files on a regular basis.

1) A. cruel　　B. crucial
2) A. antivirus　　B. virus
3) A. identity　　B. identify
4) A. removal　　B. remove

5) A. decrease　　B. increase

6) A. Because　　B. Because of

3. Please read the conversation below and learn how to raise a question and how to solve each problem.

黑客 hacker	未经许可的 unauthorized	登录 access
娱乐 fun	挑战 challenge	犯人，罪犯 criminal
惯例 convention	孤立的，隔离的 isolated	公平的 fair
无辜的 innocent	聪明的 smart	捉弄某人 playing tricks on sb.
积极的 positive	影响 influence	

A：鲍伯，你知道黑客吗？

B：知道。他们未经许可就登录他人的计算机系统来娱乐和挑战自己。他们都是罪犯。

A：你真的这样看待他们吗？实际上，黑客们也有自己的文化和惯例。

B：真的吗？但那也不意味着黑客对我们的社会有益，我们也不可能和他们成为朋友。

A：这也就是为什么黑客们被现实世界所孤立。

B：这样说是不公平的，他们伤害了那么多无辜的人。

A：黑客是非常聪明的。他们只是需要走出来与大家谈论一下系统安全问题。

B：你的意思是说他们可以帮助人们解决一些计算机安全问题而不是捉弄人们。

A：是的，他们可以对社会产生积极的影响。我想只有那样黑客文化才能被人们所了解。

A: Do you know hackers, Bob?

B: Yes, I think they are people who gain unauthorized access to a computer system for fun and challenge. They are criminals.

A: Is that what you really think of them? Actually hackers have their own culture and they have hacker conventions.

B: Really? But that still does not mean that they are good to our society and we can make friends with them.

A: That's why they are so isolated from the real world.

B: It's not fair to say so when they are hurting so many innocent people.

A: Well, hackers are pretty smart people. They just need to come out and talk to people about system security problems.

B: You mean that they can help people with security problems instead of playing tricks on them.

A: Yes. They can have a positive influence on society. I think that's really the only way the hacking culture can be understood.

4. Oral Practice

Sally 打开了一封匿名电子邮件，她的计算机因而中病毒导致系统崩溃，于是她请 John 来帮她修理。

Key Words

谢天谢地 thank goodness	别紧张，别着急 take it easy
发生故障，崩溃 break down	匿名邮件 anonymous email
好奇的 curious	抵制诱惑 resist the temptation
病毒 virus	警告 warn
冒险做某事 risk doing	更新 update
杀毒软件，防毒软件 antivirus software	

Part 2 Reading

Text

Computer Security

Why should people care about computer security?

We use computers for so many things both at home and at work, such as shopping, communicating with others through chat programs or doing business. For at home, although you may not consider the information on your computer "top secret", you still probably do not want strangers to read your personal information, to send forged email from your computer, or to corrupt the data on your computer. For at work, it is estimated that companies and organizations in America lose over $60 billion a year from computer crimes. Therefore, people pay close attention to computer security now.

Q1 What can we use computers to do?

Q2 How much money do companies and organizations in America lose a year from computer crimes?

What is computer security?

Computer security is the process of preventing and detecting unauthorized use of the computer and keeping the information and the computer itself from damaging.

Who would want to break into your computer?

Generally speaking, they are either employees, outside users (such as, company clients), hackers[1], crackers[2], organized crime members, or terrorists.

What are the forms of computer crime?

The forms of computer crime can be the creation of malicious programs, denial of service attacks, Internet scams, theft, and data manipulation, and so on.

Malicious programs are called malware, which is short for malicious software. They are specially designed to damage or disrupt a computer system. The three most common types of malware are viruses, worms[3] and Trojan horses[4]. A denial of service attack attempts to slow down or stop a computer system or network by flooding a computer or network with requests for information and data. An Internet scam is simply a scam using the Internet. Theft can take many forms. Thieves steal equipment and programs etc., such as clients information. Data manipulation is getting into someone's computer network and leaving a prankster's message. Manipulation seems harmless, it may cause great anxiety and wasted time among network users. And computer systems and data may also be destroyed by natural hazards, such as fires, flood, wind, hurricanes, tornadoes and earthquakes and so on.

How to protect computer security?

The major measures to protect computer security are encryption, restricting access, anticipating disasters and backing up data.

New Words

forge [fɔːdʒ]	v. 仿造，伪造，锻造（金属），打（铁） e. g. Yahoo tries to forge a new service to rival Google's. 雅虎试图打造一项新的服务来与谷歌竞争。
corrupt [kə'rʌpt]	v. 堕落，腐化，腐烂
detect [di'tekt]	vt. 察觉，发现，探测

unauthorized [ʌn'ɔːθəraizd]	*a*. 非法的，未被授权的，独断的 e. g. Although a firewall permits normal traffic between the outside world and the Web server, it doesn't allow authorized users outside the firewall to access the content of the Web site. 虽然防火墙允许外部世界与网站服务器之间的正常流量，但它不允许防火墙以外未经授权的用户访问网站上的内容。
hacker ['hækə]	*n*. 计算机黑客
cracker ['krækə]	*n*. 解密高手，非法破译者
terrorist ['terərist]	*n*. 恐怖主义者，恐怖分子
malicious [mə'liʃəs]	*a*. 恶意的，恶毒的，蓄意的，怀恨的 e. g. A file that contains malicious programming instructions could damage or otherwise compromise the contents of your computer. 一个包含恶意编程指令的文件可能会损坏或泄露你计算机的内容。
denial [di'naiəl]	*n*. 否认，拒绝，节制，背弃
scam [skæm]	*n*. 骗局，诡计，故事 *v*. 欺诈，诓骗 e. g. A phishing scam is a kind of crime that uses email to trick people into providing financial or other personal information. 网络钓鱼诈骗是一种利用电子邮件骗取人们财务和其他个人信息的犯罪行为。
theft [θeft]	*n*. 盗窃，偷，赃物
manipulation [məˌnipju'leiʃən]	*n*. 操纵，操作，处理，篡改
malware ['maːlˌweə(r)]	*n*. 恶意软件 e. g. Fake malware warnings have become a common method of luring innocent users into buying illegitimate software. 虚假恶意软件警告已经成为引诱无辜用户购买非法软件的通用手段。
disrupt [dis'rʌpt]	*vt*. 破坏 *a*. 分裂的，中断的
flood [flʌd]	*v*. 淹没，充满，溢出

fraudulent ['frɔːdjulənt]	a. 欺骗性的，不正的
deceptive [di'septiv]	a. 欺诈的，迷惑的，虚伪的
trick [trik]	v. 欺骗，哄骗，戏弄
prankster ['præŋkstə]	n. 爱开玩笑的人，顽皮的人，恶作剧的人
tornado [tɔː'neidəu]	n. 龙卷风，旋风，暴风，大雷雨
encryption [in'kripʃən]	n. 加密，加密术 e. g. Database encryption is a core subject in the field of information security. 数据库加密是信息安全领域研究的一个核心课题。
anticipate [æn'tisipeit]	vt. 预期，预测，占先，抢先

Key Terms

denial of service attack 拒绝服务攻击，阻断服务攻击

data manipulation 数据操纵

restricting access 访问限制

backing up data 备查资料，备份数据

Useful Expressions

be short for 是……的简称，是……的缩写

e. g. ACT used to be short for American College Test. ACT 过去是美国大学入学考试的缩写。

Notes

1. hacker

   程序设计迷，常指非法侵入他人计算机程序或网络并获取或窜改信息的人（黑客）。

2. cracker

   计算机解密者，常指非法破译者。

3. worm

   “蠕虫”病毒，一种破坏力非常强的恶意计算机程序。

4. Trojan horse

特洛伊木马程序或病毒，常用于控制他人计算机或盗取信息。

Exercises

1. Match each of the following terms with its Chinese equivalent.

1) forge	a. 恐怖分子
2) unauthorized	b. 危险
3) hacker	c. 恶意软件
4) cracker	d. 加密
5) terrorist	e. 解密高手
6) scam	f. 非法的
7) malware	g. 计算机黑客
8) hazard	h. 预测
9) encryption	i. 伪造
10) anticipate	j. 诈骗

2. Recognize the following abbreviations by matching them with their corresponding full names and translate them into Chinese.

1) NFS ______________	a. service set identifier
2) FTP ______________	b. Wi-Fi protected access
3) TCSEC ______________	c. file transfer protocols
4) WPA ______________	d. trusted computer system evaluation criteria
5) SSID ______________	e. network file system

3. Complete the following sentences by translating the Chinese in the brackets.

1) People pay close attention to ______________ (计算机安全) now.

2) Computer security is the process of ______________ (阻止和检测) unauthorized use of the computer and keeping the information and the computer itself from damaging.

3) The form of computer crime can be the creation of ______________ (恶意程序).

4) Computer systems and data may also be destroyed by ______________ (自然灾害).

5) The major measures to protect computer security are ______________ (加密、限制访问), anticipating disasters and backing up data.

4. Choose the best answer for each blank.

1) ________ infected computer may lose its data. (2008 年下半年程序员考试试题)

A. File　　B. Data base　　C. Virus　　D. Program

2) Most ________ measures involve data encryption and password. (2007 下半年程序员考试试题)

A. security　　B. hardware　　C. display　　D. program

3) The ________ in e-mail messages has affected almost every computer around the world and has caused the damage of up to US $ 1 billion in North America. (2007 上半年程序员考试试题)

A. illness　　B. virus　　C. weakness　　D. attachment

4) One of the basic rules of computer security is to change your ________ regularly. (2007 上半年程序员考试试题)

A. name　　B. computer　　C. device　　D. password

5) Firewall is a ________ mechanism used by organizations to protect their LANs from Internet. (2006 上半年程序员考试试题)

A. reliable　　B. stable　　C. peaceful　　D. security

Supplementary Reading

What Is a Firewall?

A firewall is a software program or a piece of hardware that helps screen out hackers, viruses, and worms that try to reach your computer over the Internet.

Three basic types of firewalls are available for you to choose from: software firewalls, hardware routers, wireless routers. Software firewalls are a good choice for single computers, and they work well with Windows system. (Windows 7 and Windows XP both have a built-in firewall, so an additional firewall is not necessary.) Hardware routers are a good choice for home networks that will connect to the Internet. If you have or plan to use a wireless network, you need a wireless router.

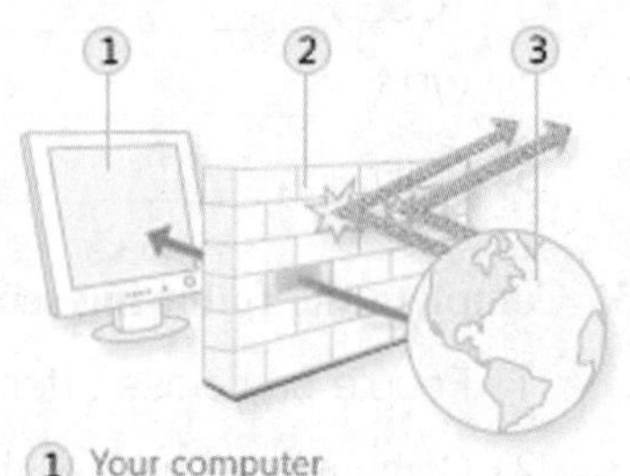

To determine which type of firewall is best for you, answer the following questions and record your answers: How many computers will use the firewall? What operating system do you use? This might be a version of Microsoft Windows, Apple Macintosh, or Linux. That's it. You are now ready to think about what type of firewall you want to use. There are several options, each with its own pros and cons. Here are some examples:

If you use a computer at home, the most effective and important first step you can take to help protect your computer is to turn on a firewall. Windows 7 and Windows XP SP2 or higher have a firewall built-in and turned on by default.

If you have more than one computer connected in the home, or if you have a small-office network, it is important to protect every computer. You should have a hardware firewall (such as a router) to protect your network, but you should also use a software firewall on each computer to help prevent the spread of a virus in your network if one of the computers becomes infected.

If your computer is part of a business, school, or other organizational network, you should follow the policy established by the network administrator.

New Words and Expressions

built-in [ˈbiltˌin] *a.* 嵌入的，固定的 *n.* 内置

version [ˈvəːʃən] *n.* (软件) 版本

option [ˈɔpʃən] *n.* 选项，选择权，买卖的特权

default [diˈfɔːlt] *n.* 违约，缺席，缺乏，系统默认值

infect [inˈfekt] *vt.* 感染，传染

screen out 筛选出

firewall software 防火墙软件

hardware routers 硬件路由器

wireless routers 无线路由器

pros and cons 正反两方面，赞成者和反对者，有利有弊

network administrator 网络管理员

Reading Comprehension

Read the following statements below and decide if they are true (T) or false (F) according to the passage you have just read.

1) A firewall is a software program or a piece of hardware that helps screen out hackers, viruses, and worms. (　　)
2) Two basic types of firewalls are available for you to choose from. (　　)
3) Software firewalls are a good choice for single computers. (　　)
4) If you use a computer at home, the most effective and important first step you can take to help protect your computer is to turn off a firewall. (　　)
5) If your computer is part of a business, school, or other organizational network, you should follow the policy established by the network administrator. (　　)

Part 3 Screen English

网络操作中的常见错误（2）

提示信息	相应含义
You can't log on as an anonymous user	你不能作为匿名用户登录
Permission denied	拒绝访问
NNTP① server error	新闻组服务器错误
Too many connections—try again later	连接太多，等一会再试
Receive and send data error	接收或发送数据错误
Connect server fail	连接服务器失败
Can't resolve server address	不能解析服务器地址
The actual size of downloading file is unknown	无法确定正在下载的文件大小
Warning: The server says 553 sorry, that domain isn't in my list of allowed reception	警告：服务器通知发生553号错误。SMTP不接受你要发送的邮件地址。

注①：NNTP（Network News Transfer Protocol）指网络新闻传输协议。

Quotation

Life is not divided into semesters. You don't get summers off and very few employers are interested in helping you find yourself. Do that on your own time.

——Bill Gates

生活不分学期。你并没有暑假可以休息，也没有几位雇主乐于帮你发现自我。自己找时间做吧。

Your school may have done away with winners and losers, but life has not. In some schools they have abolished failing grades; they'll give you as many times as you want to get the right answer. This doesn't bear the slightest resemblance to anything in real life.

——Bill Gates

你的学校也许已经不再分优等生和劣等生，但生活却仍在作出类似区分。在某些学校已经废除不及格分；只要你想找到正确答案，学校就会给你无数的机会。这和现实生活中的任何事情没有一点相似之处。

Key to Exercises

Listening & Speaking

1. 1) software　2) interfere　3) delete　4) spread　5) disk
   6) attachments　7) images　8) hidden

2. 1) B 2) A 3) B 4) B 5) A 6) A

Exercises

1. 1) i 2) f 3) g 4) e 5) a 6) j 7) c 8) b 9) d 10) h
2. 1) e 网络文件系统 2) c 文件传输协议 3) d 可信任的计算机系统评价标准
   4) b Wi-Fi 保护访问 5) a 服务集标识
3. 1) computer security
   2) preventing and detecting
   3) malicious programs
   4) natural hazards
   5) encryption, restricting access
4. 1) C 2) A 3) B 4) D 5) D

Comprehension

1) T 2) F 3) T 4) F 5) T

附　录　计算机专业英语词汇表

A

active-matrix 主动矩阵

adapter card 适配卡

advanced application 高级应用

analytical graph 分析图表

analyze 分析

animations 动画

application software 应用软件

arithmetic operation 算术运算

audio-output device 音频输出设备

access time 存取时间

access 存取

accuracy 准确性

ad network cookies 广告网络信息记录软件

administrator 管理员

add-on 插件

address 地址

agent 代理

analog signal 模拟信号

applet 程序

asynchronous communications port 异步通信端口

attachment 附件

B

bar code 条形码

bar code reader 条形码读卡器

basic application 基础程序

binary coding scheme 二进制编码方案

binary system 二进制系统

bit 比特

browser 浏览器

backup tape cartridge unit 备份磁带盒单元

bandwidth 带宽

bluetooth 蓝牙

broadband 宽带

business to business 企业对企业

business to consumer 企业对消费者

bus 总线

C

cable 电缆

cell 单元箱

chain printer 链式打印机

character and recognition device 字符标识识别设备

chart 图表

chassis 支架

chip 芯片

clarity 清晰度

closed architecture 封闭式体系结构

column 列

combination key 结合键

computer competency 计算机能力

connectivity 连接，节点

continuous speech recognition system 连续语言识别系统

control unit 操纵单元

cordless/wireless mouse 无线鼠标

cable modems 有线调制解调器

carpal tunnel syndrome 腕管综合征

CD-ROM 只读光盘

CD-RW 可重写光盘
CD-R 可记录压缩光盘
channel 信道
chat group 谈话群组
chlorofluorocarbons（cfcs）氯氟甲烷
client 客户端
coaxial cable 同轴电缆
cold site 冷网站
commerce server 商业服务器
communication channel 信道
communication system 信息系统
compact disc rewritable 可重写光盘
compact disc 光盘
computer abuse amendments act of 1994 计算机滥用法案（1994）
computer crime 计算机犯罪
computer ethics 计算机道德
computer fraud and abuse act of 1986 计算机欺诈和滥用法案（1986）
computer matching and privacy protection act of 1988 计算机查找和隐私保护法案（1988）
computer network 计算机网络
computer support specialist 计算机支持专家
computer technician 计算机技术人员
computer trainer 计算机教师
connection device 连接设备
consumer-to-consumer 个人对个人
cookies-cutter program 信息记录截取程序
cookies 信息记录程序
cracker 解密高手，非法破译者
cumulative trauma disorder 积累性损伤错乱
cyber cash 电子货币
cyberspace 网络空间
cynic 愤世嫉俗者
chart 图表
closed architecture 封闭式体系结构

D

database 数据库
database file 数据库文件
database manager 数据库管理
data bus 数据总线
data projector 数码放映机
desktop system unit 台式计算机系统单元
destination file 目标文件
digital camera 数码照相机
digital notebook 数字笔记本
digital video camera 数码摄影机
discrete-speech recognition system 不连续语言识别系统
document 文档
document file 文档文件
dot-matrix printer 点阵式打印机
dual-scan monitor 双向扫描显示器
dumb terminal 非智能终端
data security 数据安全
data transmission specification 数据传输说明
database administrator 数据库管理员
dataplay 数字播放器
demodulation 解调
denial of service attack 拒绝服务攻击
dial-up service 拨号服务
digital cash 数字现金
digital signal 数字信号
digital subscriber line 数字用户线路
digital versatile disc 数字化通用光盘
digital video disc 数字化视频光盘
direct access 直接存取
directory search 目录搜索

disaster recovery plan 灾难恢复计划
disk caching 磁盘驱动器高速缓存
diskette 磁盘
disk 磁碟
distributed data processing system 分部数据处理系统
distributed processing 分布处理
domain code 域代码
download 下载
DVD-R 可写 DVD
DVD-RAM DVD 随机存取器
DVD-ROM 只读 DVD

E

e-book 电子阅读器
expansion card 扩展卡
end user 终端用户
e-cash 电子现金
e-commerce 电子商务
electronic cash 电子现金
electronic commerce 电子商务
electronic communications privacy act of 1986 电子通信隐私法案（1986）
encrypt 加密
energy star 能源之星
enterprise computing 企业计算化
environment 环境
erasable optical disk 可擦除式光盘
ergonomics 人类工程学
ethics 道德规范
external modem 外置调制解调器
extranet 企业外部网

F

fax machine 传真机
field 域
find 搜索
firewire port 火线端口
firmware 固件
flash RAM 闪存
flatbed scanner 台式扫描器
flat-panel monitor 纯平显示器
floppy disk 软盘
formatting toolbar 格式化工具条
formula 公式
function 函数
fair credit reporting act of 1970 公平信用报告法案（1970）
fiber-optic cable 光纤电缆
file compression 文件压缩
file decompression 文件解压缩
filter 过滤
firewall 防火墙
fixed disk 固定硬盘
flash memory 闪存
flexible disk 可折叠磁盘
floppies 磁盘
floppy-disk cartridge 磁盘盒
formatting 格式化
freedom of information act of 1970 信息自由法案（1970）
frequency 频率
frustrated 受挫折
full-duplex communication 全双通通信

G

general-purpose application 通用运用程序
gigahertz 千兆赫
graphic tablet 绘图板
green PC 绿色个人计算机
group 分组

H

handheld computer 手提电脑
hard copy 硬拷贝

hard disk 硬盘
hardware 硬件
help 帮助
host computer 主机
home page 主页
hyperlink 超链接
hacker 黑客
half-duplex communication 半双通通信
hard-disk cartridge 硬盘盒
hard-disk pack 硬盘组
head crash 磁头碰撞
header 标题
help desk specialist 帮助办公专家
helper application 帮助软件
hierarchical network 层次型网络
history file 历史文件
hit 匹配记录
horizontal portal 横向用户
hot site 热网站
hybrid network 混合网络

I

image capturing device 图像获取设备
information technology 信息技术
ink-jet printer 喷墨式打印机
integrated package 综合性组件
intelligent terminal 智能终端设备
intergrated circuit 集成电路
implement 实现接口
interface card 接口卡
internal modem 内部调制解调器
internet telephony 网络电话
internet terminal 互联网终端
identification 识别
i-drive 网络硬盘驱动器
illusion of anonymity 匿名幻想
index search 索引搜索
information pusher 信息推送器
initializing 初始化
instant messaging 即时信息
internal hard disk 内置硬盘
internet hard drive 网络硬盘驱动器
intranet 内联网

J

joystick 操纵杆

K

keyword search 关键字搜索

L

laser printer 激光打印机
layout file 版式文件
light pen 光笔
locate 定位
logical operation 逻辑运算
land 凸面
line of sight communication 视影通信
low bandwidth 低带宽
lurking 潜伏

M

main board 主板
mark sensing 标志检测
mechanical mouse 机械鼠标
memory 内存
menu 菜单
menu bar 菜单栏
microprocessor 微处理器
microsecond 微秒
modem card 调制解调器

monitor 显示器
motherboard 主板
mouse 鼠标
multifunctional device 多功能设备
magnetic tape reel 磁带卷
magnetic tape streamer 磁带条
mailing list 邮件列表
medium band 媒质带宽
metasearch engine 整合搜索引擎
microwave 微波
modem 调制解调器
modulation 调制

N

net PC 网络计算机
network adapter card 网卡
network personal computer 网络个人计算机
network terminal 网络终端
notebook computer 便携式计算机
notebook system unit 便携式计算机系统单元
numeric entry 数字输入
national information infrastructure protection act of 1996 国家信息基础设施保护法案（1996）
national service provider 全国性服务供应商
network architecture 网络体系结构
network bridge 网桥
network gateway 网关
network manager 网络管理员
newsgroup 新闻组
no electronic theft act of 1997 无电子盗窃法（1997）
node 节点
nonvolatile storage 非易失性存储

O

object embedding 对象嵌入
object linking 目标链接
open architecture 开放式体系结构
operation system 操作系统
optical disk 光盘
optical mouse 光电鼠标
optical scanner 光电扫描仪
outline 大纲
off-line browser 离线浏览器
online storage 联机存储

P

palmtop computer 掌上电脑
parallel port 并行端口
passive-matrix 被动矩阵
PC card 个人计算机卡
personal laser printer 个人激光打印机
personal video recorder card 个人视频记录卡
photo printer 照片打印机
pixel 像素
platform scanner 平板式扫描仪
plotter 绘图仪
plug and play 即插即用
plug-in board 插件板
pointer 指示器
pointing stick 指示棍
port 端口
portable scanner 便携式扫描仪
presentation file 演示文稿
presentation graphics 电子文稿程序
primary storage 主存
procedure 规程
processor 处理器
programming control language 程序控制语言
packet 数据包

parallel data transmission 平行数据传输
peer-to-peer network system 点对点网络系统
person-person auction site 个人对个人拍卖站点
physical security 物理安全
pit 凹面
plug-in 插件程序
privacy 隐私权
proactive 主动地
programmer 程序员
protocol 协议
provider 供应商
project 项目，工程
proxy server 代理服务器
pull product 推取程序
push product 推送程序

R

RAM cache 随机高速缓冲器
range 范围
record 记录
relational database 关系数据库
replace 替换
resolution 分辨率
row 行
read-only 只读
reformatting 重组
regional service provider 区域性服务供应商
repetitive motion injury 反复性动作损伤
reverse directory 反向目录
right to financial privacy act of 1979　财产隐私法案（1979）
ring network 环形网

S

scanner 扫描器
search 查找
secondary storage device 辅助存储设备
semiconductor 半导体
serial port 串行端口
server 服务器
shared laser printer 共享激光打印机
sheet 表格
silicon chip 硅片
slot 插槽
smart card 智能卡
soft copy 软拷贝
software suite 软件协议
sorting 排序
source file 源文件
special-purpose application 专用文件
spreadsheet 电子数据表
standard toolbar 标准工具栏
supercomputer 巨型计算机
system 系统
systemcabinet 系统箱
system clock 时钟
system software 系统软件
satellite/air connection service 卫星无线连接服务
search engine 搜索引擎
search provider 搜索供应方
search service 搜索服务
sector 扇区
security 安全
sending and receiving device 发送接收设备
sequential access 顺序存取
serial data transmission 单向通信
signature line 签名档
snoopware 监控软件
software copyright act of 1980　软件版权

法案（1980）
software piracy 软件保密
solid-state storage 固态存储器
specialized search engine 专用搜索引擎
spider 网页爬虫
spike 峰值
star network 星形网
strategy 方案
subject 主题
subscription address 预定地址
superdisk 超级磁盘
surfing 网上冲浪
surge protector 浪涌保护器
system analyst 系统分析师

T

table 表
telephony 电话学
television board 电视扩展卡
terminal 终端
template 模板
text entry 文本输入
thermal printer 热印刷
thin client 瘦客户端
toggle key 触发键
toolbar 工具栏
touch screen 触摸屏
trackball 追踪球
TV tuner card 电视调谐卡
two-state system 双状态系统
technical writer 技术协作者
technostress 科技压力
telnet 远程终端协议
time-sharing system 分时系统
topology 拓扑结构
track 磁道
traditional cookies 传统的信息记录程序
twisted pair 双绞

U

unicode 统一字符标准
upload 上传
Usenet 世界性新闻组网络

V

virtual memory 虚拟内存
video display screen 视频显示屏
voice recognition system 声音识别系统
vertical portal 纵向门户
video privacy protection act of 1988
视频隐私权保护法案（1988）
virus checker 病毒检测程序
virus 病毒
voiceband 音频带宽
volatile storage 易失性存储
voltage surge 电涌

W

wand reader 条形码读入
Web 网络
Web appliance 万维网设备
Web page 网页
Web site address 网络地址
Web terminal 万维网终端
webcam 网络摄像头
what-if analysis 假定分析
wireless revolution 无线革命
word length 字长
word processing 文字处理
word wrap 自动换行
worksheet file 工作文件
Web auction 网上拍卖

Web broadcaster 网络广播

Web portal 门户网站

Web site 网站

Web storefront creation package 网上商店创建包

Web storefront 网上商店

Web utility 网上应用程序

Web-downloading utility 网络下载应用程序

webmaster 站点管理员

Web 万维网

wireless modem 无线调制解调器

wireless service provider 无线服务供应商

World Wide Web 万维网

worm 蠕虫病毒

write-protect notch 写保护缺口

参 考 文 献

[1] Charles S Parker，Deborah Morley，Brett Miketta. 计算机专业英语 [M]. 北京：科学出版社，2003.

[2] 陈枫艳，陈志峰. 计算机英语 [M]. 北京：清华大学出版社，北京交通大学出版社，2010.

[3] 陈秋劲. 计算机英语教程 [M]. 北京：高等教育出版社，2008.

[4] 任军战. 计算机英语 [M]. 北京：外语教学与研究出版社，2008.

[5] 司炳月. IT 精英职场英语口语 [M]. 大连：大连理工大学出版社，2010.

[6] 苏慧明. 计算机英语 [M]. 北京：高等教育出版社，2008.

[7] 盛时竹，丁秀芹，殷树友. 计算机专业英语 [M]. 北京：清华大学出版社，2006.

[8] 温涛，张翼，杨毅. IT 行业英语 [M]. 北京：清华大学出版社，2008.

[9] 王翔. 实用 IT 英语 [M]. 天津：天津大学出版社，2009.